Nontraditional Students and Community Colleges

Nontraditional Students and Community Colleges

The Conflict of Justice and Neoliberalism

John S. Levin

macmillan

Nontraditional Students and Community Colleges: The Conflict of
Justice and Neoliberalism
Copyright © John S. Levin, 2007.

All rights reserved. No part of this book may be used or reproduced in any manner
whatsoever without written permission except in the case of brief quotations
embodied in critical articles or reviews.

First published in 2007 by
PALGRAVE MACMILLAN™
175 Fifth Avenue, New York, N.Y. 10010 and
Houndmills, Basingstoke, Hampshire, England RG21 6XS.
Companies and representatives throughout the world.

PALGRAVE MACMILLAN is the global academic imprint of the Palgrave
Macmillan division of St. Martin's Press, LLC and of Palgrave Macmillan Ltd.
Macmillan® is a registered trademark in the United States, United Kingdom, and
other countries. Palgrave is a registered trademark in the European Union and
other countries.

ISBN-13: 978-1-4039-7010-7
ISBN-10: 1-4039-7010-6

Library of Congress Cataloging-in-Publication Data

A catalogue record of the book is available from the British Library.

Design by Scribe Inc.

First edition: August 2007

10 9 8 7 6 5 4 3 2 1

Printed in the United States of America.

Contents

Acknowledgments

I count myself fortunate that there are many community college administrators, faculty, staff, and students who generously gave of themselves to my research. Considerable assistance was provided to me during my field studies, and among the many I name a few. Jack Oharah, Jerrilee Mosier, and Tasha King at Edmonds Community College; Ken Meier and Karen Sallee at Bakersfield College; Christine Johnson and Wanda Underwood at Community College of Denver; Don Reichard and Talbert Myers at Johnston Community College; Martha Williams and Alisa Nagler at Wake Technical Community College; John Means at El Camino Community College; Antonio Perez and Richard Hasselbach at the Borough of Manhattan Community College; David Hardy at Mountain View College; Jana Kooi at Pima College; and Susan Kater at GateWay Community College. There are many others and they will know that I am thankful for their help.

I am grateful for the support from a number of sources for my work. The Lumina Foundation for Education provided financial support for the research that is the backbone of this book. The Joseph D. Moore endowed professorship at North Carolina State University gave me considerable latitude as a scholar, which included financial aid for researchers and graduate students. My publisher, Palgrave Macmillan, continues to acknowledge my worth, and my editor, Amanda Johnson Moon, and her staff are convincingly positive about my books. For a solitary researcher/author, this attitude is nourishing.

I was assisted in my research by several worthy graduate students at North Carolina State University, including Jerrid Freeman, Erin Robinson, Laura Gonzalez, Jennifer Hildreth, Morgan Murray, Candice Fisher, and David Frye. Jerrid's doctoral dissertation relied upon this project, and his work on the project moved me forward and helped in the development of Chapter 5. Laura read through drafts of chapters and provided me with some fine questions that, once answered, strengthened the book. Supporting me as always in my travels and financial activities was Dawn Crotty. Amy Metcalfe, formerly a doctoral student at the University of Arizona, and presently an assistant professor at the University of British Columbia, initiated my journey into sophisticated technologically assisted data analysis, and she provided me with a first run of computer-assisted analysis. My colleagues at North Carolina State University—particularly Duane Akroyd, Audrey Jaeger, Crystal Muhammad, Leila Gonzalez Sullivan, and Marvin Titus—were always positive about my research and never gave any signals that my many travels to sites were a burden to them. Carol Kasworm, department head of Adult and Higher Education, and Kay Moore, dean of the College of Education, at North Carolina State University, always spoke in grandiose terms about my research, proving that flattery goes a long way.

Key to the production of this book were three colleagues—Marilyn Amey of Michigan State University, Estela Bensimon of the University of Southern California, and Ken Meier of Butte College. They willingly, without arm-twisting, agreed to review the manuscript for this book and offer both editorial and scholarly advice. Their careful reading and thoughtful advice improved this book and gave me confidence in letting the book move on to the public.

Of special significance to me were the many hours of telephone conversations with Brian Pusser of the University of Virginia. Brian and I would debate my constructs repeatedly, and Brian was a provocateur for my research and my thinking about the deeper issues. Indeed, without Brian's grantsmanship, there would have been no funding from Lumina, and the many hours Brian devoted to the project, whose members included David Breneman, Kay Kohl, Bruce Gansneder, John Milam, and Sarah Turner, was time given to the good of us all.

Going further back in time, I recognized during this book writing that my early career experiences at Douglas College, Kwantlen College, and North Island College in British Columbia, Canada, shaped how I view community colleges, particularly their students. I recalled during my review of interview transcripts with faculty, administrators, and students—some 180 of these—that there was considerable similarity between the perceptions and views emanating from these colleges and the memories I possessed from my experiences as a college instructor and administrator.

Almost neglected by me in my continued focus upon research over the past several years are my family, including adult children and grandchildren—Tracy, D'Arcy, Simon, Jeremy, Mathew, Hannah, Sophia, Ethan, Fraser, James, and Kai—who have accepted me as I am and seem to be always delighted to hear from me and see me. As well, my three daughters-in-law—Anita, Susan, and Lindsay—were interested in my academic exploits. Of all my publications, this book is the one that children and grandchildren will likely understand and find useful, or at least amusing, and someday that will include those who do not yet read. My wife, Lee, lives with my research because she lives with me, both literally and figuratively, and she not only copes with my field work absences but also responds intelligently when I need to check out an idea or interpretation.

This project was a new direction for me, and it inspired me to move further along this path by developing this story for another medium—film. This book, then, has been an impetus for me to modify traditional social science research into another form so that dissemination of my findings could occur in other forms, particularly the visual and auditory. My new partner in this venture is my son, Jeremy Levin, whose patience is exemplary and whose artistic and technical talents have helped to shape the research into film. As this film project moved forward, the filming of new students influenced the writing of this book as earlier ideas became either reinforced or modified by my experiences.

Finally, John Rawls' theory of justice—that ever-present chord that rings through this book—evoked in me memories of my father, Norman Levin, who as a criminal defense lawyer conducted dialogues with me when I was eight and nine about justice, inculcating a sentiment in me for the disadvantaged, the forgotten, and the maligned in our society. Although Norman did not accompany me into my youth, as he died far too soon, he would have, I am certain, derived much satisfaction from my use of Rawls. I am grateful to my father for accompanying me in my imagination on my journey of writing this book.

Introduction

The Themes of Justice and Neoliberalism

Although the United States is the acknowledged world leader in a multitude of endeavors, fields, and performance measures—including the world's wealthiest nation—the country lags behind other countries in the more humanitarian domains. The United States has one of the weakest sets of social and economic welfare programs of postindustrial states; it has the highest proportion of its population living in poverty among those states; and it has unparalleled social and economic inequities based upon race.[1] This inequity certainly applies to education and higher education as well.[2] With the emergence of global economic competition as a defining factor in U.S. society beginning in the 1980s,[3] including efforts to increase productivity and efficiency through the use of labor-saving technologies,[4] the development of new economic classes was underway.[5] In globalization terms, these classes were bimodal: the "haves" and the "have-nots."

The same pattern is evident in higher education institutions. Economic competition has led not only to stratification among institutions but also within institutions, a condition evident in all types of institutions.[6] The "have" institutions and the "have" programs are those with wealth, prestige, and impact: Their faculty garner funding for themselves and their institutions through business and industry partnerships, inventions, patents, and research grants; their students obtain good jobs or places in selective graduate or professional programs; and their donors, many former students of wealth, sustain the enterprise. Athletics in the form of big-time male sports, while not in actuality a profit center, keeps the alumni as supporters of the institution and ultimately donors.[7] Clearly, higher education has privileged some institutions and programs as well as organizational participants, including administrators, faculty, and students. Indeed, students as the recipients of educational services are privileged under certain conditions: if they attend the elite or wealthy colleges, enroll in the prestigious or marketable programs, and perform in the expected manner. These students are commonly students who have entered college directly from high school, possessing high or moderately high socioeconomic standing and the cultural and social capital, as well as the grades and test scores, that will ensure them of acceptable college performance.[8]

The "have-not" or low salient institutions and programs include those colleges without wealth, prestige, and social and economic impact and those programs that are undervalued. For low prestige institutions, such as community colleges, undervalued programs are those with little overt connection to either university transfer or occupations, such as English as a second language (ESL), Basic Skills, Adult High School, and vocational certificate programs. Compared to prominent four-year colleges and universities, community colleges can be viewed as "have-not" institutions, and all of their students, subject to negative judgments, are viewed as on the periphery of middle-class status. Clearly, earnings from the result of baccalaureate degree attainment indicate that nonbaccalaureate institutions and their students are of lesser value than four-year colleges.[9] This is not to suggest that all four-year colleges and their students are privileged. Indeed, the majority of four-year colleges compared to the minority of elite colleges are without wealth, prestige, and influence. Nonetheless, the stratification of institutions places community colleges, tribal colleges, and technical institutions at the bottom of the hierarchy.

The hierarchy of institutions is well matched to the hierarchy of student profiles. At one end of the spectrum are elite private colleges and research universities serving one type of student and, at the other end, the community college, including technical colleges, serving another type of student. This hierarchy includes segmented student populations, with new or nontraditional students found at the lower end of the spectrum. For all the talk of nontraditional students—such as those over twenty-four years of age and those who are first-generation attendees—in higher education, the bulk of these nontraditional students are enrolled in community colleges. A conservative estimate based upon national data suggests that 65 percent of nontraditional students enrolled in credit undergraduate programs at colleges and universities attend community colleges.[10] Of the 5.6 million credit-seeking community college students in 2000, close to 90 percent have one characteristic, such as delayed postsecondary enrollment or part-time attendance, that would classify them as nontraditional.[11] In addition, there are an estimated 5 million community college students enrolled in noncredit courses and programs who could be classified as nontraditional.[12]

The problems of privilege—those who are excluded and neglected—directly define the community college and its students. They are outsiders to privilege, on the periphery of prosperity, and at the bottom strata of prestige for institutions and students. In addition, the political economy favors neoliberal values and goals, with competition as the norm.[13] In such an environment, those without privilege are unjustly served and their rewards are few. In the field of higher education, these institutions and their students are surely "have-nots." Furthermore, the issues of individual identities, where some are positively assessed and judged and others are negatively judged, are connected to institutional affiliation.[14] Community colleges viewed as institutions for specific populations—disadvantaged students, minority students, low economic status students—indicates that these populations' identities are tied to the institution's identity and to its restraints.

A salient example of those without privilege, both in our society and in higher education institutions, are single mothers participating in government welfare assistance programs. They are both without privilege and they are stigmatized because of their participation in social welfare.[15] Their participation in higher education is aimed at the learning-earning connection, but it is also a way for them to change their lives. Furthermore, their children become part of that rationale as they seek to provide for and model preferred outcomes for their children.[16] One of the painful ironies is that the government program that provides them with temporary financial assistance discourages education and training, and when it does support education and training it does so for the benefit of local labor markets.[17] Indeed, there is evidence that policies that put work before education do not enhance this population's upward mobility out of poverty.[18] Federal policy clearly supports an ideology where the euphemistic notion of personal responsibility takes precedence and the states, generally, follow the same general trend, with several competing ideas about the role of education for this population.[19] One specific issue is that a two-year time limit is placed upon welfare assistance, and with a one-year limit upon education, there is a disincentive for the completion of long-term programs, especially four-year degree programs.[20] This example points to the role of the state in limiting educational opportunities for a specific population, and the state's actions thwart opportunities for those whose circumstances are certainly not privileged: They are disadvantaged in the extreme. In my argument, then, these students are denied justice.

These conditions are tied to the goals and ambitions of neoliberals, who in the decades of the 1990s and 2000s have succeeded in reshaping the discourse of social welfare and entitlement, including the responsibilities of the state in providing for its citizens.[21] Proponents of neoliberalism combine a commitment to state reduction of social programs and state support for global economic markets, with economic competition as a way to increase efficiency and productivity and bring wealth to the successful.[22] Because neoliberalism is a political project aimed at institutional change,[23] postsecondary education is an arena under severe pressure to alter its historical norms of governance, organization, and mission.[24] Both its economic goals and its preferred processes of meeting those goals make neoliberalism antithetical to justice for the disadvantaged student.

Neoliberal ideology favors individual achievement and economic prosperity for individuals, relegating social benefits and progress to an assumed by-product of individual economic behaviors. Its view of fairness is unregulated markets (i.e, free trade) and equal opportunity (which governments may be required to structure and monitor) in the form of access to economic prosperity. Neoliberalism does not have an historical perspective and ignores the prior conditions of competition between and among individuals and groups. Thus, there is no assumption that some populations or individuals need advantages in order to compete fairly with others. As well, with an emphasis upon economic benefits and outcomes, neoliberalism subordinates cultural and social goals—such as community or family cohesion—treating them more like means rather than as ends.[25]

In this book, I examine if, and the extent to which, community college students receive justice both within their institution and as an outcome of their education. I determine that while individual students may receive some measure of justice, groups of students or classes of students do not. I also argue that individual action and institutional context, such as organizational culture, not policy, are responsible for the justice that is meted out to students. My findings are the result of extended interviews with administrators, faculty, and students, as well as state higher education policymakers.

By justice I mean a condition aligned with fairness and the equalizing of advantage so that prior conditions for individuals are recognized and accounted for in rights, privileges, and treatment that compensate for an individual's disadvantage. I borrow from John Rawls' concept of justice as fairness and his articulation of a well-ordered society that operates through social cooperation, under a social contract.[26] I examine the actual condition of students in community colleges—institutions that I view as components of a well-ordered society—to understand these students and to ascertain how the institution treats these students.

My work extends across the nation as I observed and interacted with students, faculty, administrators, and state-level policymakers over the period of 2002 to 2006. My travels took me to Arizona and GateWay Community College and Maricopa Community College District, and Pima Community College; to California and Bakersfield College and El Camino College; to Colorado and Community College of Denver; to New York and Borough of Manhattan Community College; to North Carolina and Johnston Community College and Wake Technical Community College; to Texas and Mountain View College and Dallas Community College District; to Washington and Edmonds Community College; and, with the help of my research assistants, to Illinois and Harry Truman College, and to Virginia and Piedmont Community College and Virginia Highlands Community College.

My meetings were in most cases extended conversations with a broad spectrum of people affiliated with community colleges. I met with students who possess various attributes, some with disabilities such as physical and mental impairment, blindness, and deafness; some with learning problems; many with Spanish as their native language; some with undocumented immigrant status in the United States; several with the hardships of life woven into their personal accounts of their educational experiences. I talked with administrators and faculty members who were, almost to a person, supporters of students and student achievement in their college education. I met with several state-level officials who were sensitive to student needs and disadvantages. Most of these officials justified the community college's purposes as connected to student advancement, and particularly to the improvement of students' lives. With such expressions of concern and compassion for disadvantaged students by state-level officials and institutional faculty and administrators, the reasonable expectation is that students at community colleges are well served.

But, in some distinction to my experiences with those who represented disadvantaged students and their supporters, I viewed institutional units and talked with institutional officials who were either engaged in projects and activities that ignored the needs of students or pressured to turn their attention to matters that may have impacted these students negatively. Simply, students were used to meet the needs of the nation-state or the state or the community, largely in the form of business and industry. Indeed, in the extreme, the community college and its students were servants of multinational corporations and government departments, such as the Department of Defense, for training both a workforce and supplying the company or the nation with skilled labor. The pressures here stemmed from the community college's assumed need to acquire resources either to subsidize a low revenue generating area of the college or to provide training to community members: These resources were not available from state government base allocations.

It might be simplistic to conclude that money—or the lack thereof— is the root of evil for community colleges: Those with fiscal resources influence the direction of these institutions, and community colleges, themselves, without sufficient fiscal resources, cannot serve all of the needs of students or communities. Insufficient resources were the common theme among community college practitioners, from faculty to college presidents to system chief executives, in my conversations. Both the behaviors of the state in underfunding higher education institutions and the seemingly rapacious appetites of these institutions to require and use fiscal resources suggest a condition of resource dependency.[27]

For community colleges, this means dependency upon resource providers— students, government agencies, local and state and federal governments, private foundations, and business and industry. Students become both commodities and consumers—sources of revenue and products to be sold.[28] As these providers increase their influence within the institution—often for self-interest—the mission and actions of the institution alter to reflect and respond to these influencers.[29] My observations during my visits to thirteen campuses and college systems provided both confirming and disconfirming evidence of the money theme and the condition of resource dependency.

The issues were more complicated, reflecting not only the institution, such as its identity as a community college with its state relationship, but also historical conditions and demographic developments.[30] The infamous "9/11" attack and its aftermath affected not just the general economic conditions of the United States, but individual institutions as well, with the Borough of Manhattan Community College as an exceptional example. Additionally, the major wave of immigration in the past two decades, including undocumented immigrants, altered not just community college programming, for example, a rise in remedial and English as a second language (ESL) programs,[31] but as well college culture in general and student services in particular as colleges adjusted to new populations.

Nontraditional Students

"Nontraditional" and "non-traditional student" are problematic terms for both scholars and practitioners. This book will work at shaping understanding of student populations in higher education so that the distinctions between "traditional" and "nontraditional" students might be understood in new ways; for example, as distinctions of privilege, contrasts between those who have advantages and those who are disadvantaged, and those whose experiences of college are categorically different from other student populations. I might accept that there are two historical groups, traditional and nontraditional, but over the past twenty years a new group of nontraditional students have begun populating colleges and it is better referred to as "new nontraditional students." Indeed, I will conclude this book in the final chapter with this group. For now, however, I will use the two terms—traditional and nontraditional—so as not to further compound the problem of definition.

Traditional students are customarily viewed from a four-year college and university perspective as the norm: students who have characteristics that place them in the mainstream of college students. Among the several defining traits of this population, their student identity—students first and other attributes second—is paramount. They are viewed as students who have continued their education from high school to college or university, thus their age at college entry is seventeen, eighteen, or nineteen. They are also viewed as full-time students, even if they are not undertaking a *de facto* full course load but *de jure* full course load. (For example, if five courses per semester is a full load, three or four courses may categorize a student as full-time.) Other characteristics are not as firmly connected to the traditional student, but they are generally viewed as their attributes. These include high school completion, second or next generation of postsecondary education attendance within a family, and English as a first language. Of course, there is a continuum of traditionality from those students who live on campus, are from middle-class backgrounds, and have clear degree aspirations to those who are commuter students, from lower-middle-class and working-class backgrounds, and who have unclear educational goals.

Thus, using this conception of traditional students, nontraditional students can be viewed as the antithesis of the traditional. However, in some institutions, such as the community college, the nontraditional student is more the norm than the traditional. In Chapter 1, the continuum of nontraditional students who are present in these settings will be discussed. To help illustrate the lack of clarity that muddies this conversation, however, we will begin with the example of adult students. This is but one of the categories that comprises the larger category of nontraditional, and the level of disagreement over a basic definition of "adult" serves as a caution.

Nontraditional Students as Adults

The understanding of the experiences of nontraditional students in higher education institutions is confounded by the conflation of the broad categories of nontraditional and adult. Considerable confusion has resulted from using the general term "adult students" in varied ways. Customarily, the literature on adult students does not delineate such matters as student demographics, institutional type, and the theoretical assumptions and purposes of the scholars and those reporting data. The understanding of adult students is further complicated by the varying definitions that researchers, educators, and other service providers give to them. The U.S. Department of Education categorizes adult education as (1) basic needs (ESL or Adult Basic Education students), (2) vocational (apprenticeship or certificate programs), or (3) educational (either job/career-related, such as credentialing programs, or nonwork-related, such as personal interest and development classes).[32] These categories include credit and noncredit offerings, community-based programs, and university or community college–based programs. The considerable breadth in these descriptions of adult education settings can include recent immigrants with minimum-wage positions who are trying to acquire basic English skills to middle-aged careerists who are trying to enhance their chances for promotion.

In addition to confusion in the literature on adult students, the problem of understanding student experiences and the institution-student intersect is exacerbated with a major body of literature's focus not upon adult students but upon adult education. Adult education is defined in mainstream scholarship and reported as any activity where the participants are adults—usually twenty-four years of age or older.[33] Yet educational data on the twenty-four-year and older category are not conceptualized coherently by the National Center for Education Statistics, which reports that those age 24 and older are automatically considered financially independent.[34] Those in the 24–29 age category are in the lowest income quartile, with 41.2 percent of the students aged 24–29 in that grouping. The highest income quartile consists of those over age 40, with 41.5 percent in that group. Both of these groups are viewed as adult students, but their conditions are unlikely to be similar.

Kopka, Schantz, and Korb, who report 1991 data, indicate that 17 percent of course-taking by adults occurred in four-year colleges, and another 20.8 percent in two-year colleges or technical/vocational schools.[35] The largest percentage (27.1 percent) was provided by business and industry. Eleven percent of courses undertaken were in a federal, state, or local government setting (including the military), and another 7.5 percent were offered by community groups. The remaining providers include private instruction, professional associations, or secondary schools. Their respondents, those people who completed the National Household Education Survey, were more likely to participate in adult education if they were already employed and had completed a bachelor's degree previously. Thus, a degree of privilege may already be present for these particular adult students. When the authors disaggregated the sample by employment status, the data indicated that unemployed persons were more likely to take classes in community colleges or vocational schools than elsewhere (49.7 percent) while employed persons were more likely to utilize their business settings (29.5 percent). This sample may typify one population of adult learners but not all populations. I was particularly fortunate to have the support of graduate student Virginia Hernandez at the University of California, Riverside, who not only did the bulk of the index for this book but also moved my thinking beyond this book by her reexamination of the original collected data.

Many of these distinctions in definition depend on the purpose and nature of those who define the adult learner. The government may use one set of criteria, research literature another, and community agencies yet another. In higher education literature, adult students are frequently described as motivated, perhaps a little "rusty," but capable and resourceful. They are assumed to seek career advancement and engage in credit-bearing programs at community colleges or universities. Carol Aslanian,[36] who examines adult students in all postsecondary educational settings, states that those in this group "lead very busy lives, juggling career and family roles. Their family incomes are higher than for most other American households. They have high levels of education as they return to college and are eager to raise those levels even further" (56).

Aslanian's population, however, is not typical of those who enroll in large numbers in community colleges in such areas as Adult High School, Basic Skills, English as a second language, and certificate programs. Furthermore, data on these populations are difficult to obtain, and data-driven assertions about adult students and their program choices are problematical. Adults pursuing a General Educational Development (GED) credential, for example, may be taking classes from a high school, from a community agency, from the noncredit segment of the community college, or as a credit-seeking community college student.[37] National figures are unavailable. Mingle, Chaloux, and Birkes note that "data are hard to obtain on enrollment in Adult Basic Education, adult secondary education, and English literacy programs for those who do not speak English. Funding comes from many different sources, and many of these programs are sponsored by private agencies that do not report their enrollments to states."[38]

As researchers endeavor to describe adult students, they may utilize different sets of defining characteristics or criteria. One characteristic is occupational level. The population traditionally conceived of as mature, with substantial work and life experiences, as well educated and economically satisfied, is a distinctly different population from that served by welfare offices, immigration specialists, and social workers. Grubb describes low-skilled workers and their various reasons for enrollment in a community college.[39] While some are experienced workers who have not been promoted and require additional courses to complete a licensure or improve their occupational standing, others are unemployed or dislocated workers who are compelled to change their life's directions. As well, there are those who face employment barriers, and this population includes welfare recipients, new immigrants, inmates released from prison, students with disabilities, or those who have been unemployed for long stretches of time. As low-skilled workers, these adults struggle with survival issues: They lack requisite social and cultural capital to thrive in a competitive labor force.[40]

"Risk factors" is another concept tied to adult students, but the details are not disaggregated by institutional type. A National Center for Education Statistics (NCES) table, "Percentage of 1999–2000 undergraduates with various risk characteristics," addresses "risk factors" for students, including part-time attendance at college, delayed enrollment, having dependents, and working while enrolled.[41] These risk factors are a component of a student attainment focus, and there is a reputed correlation between risk factors and lack of student educational attainment.[42] Students aged 24 and older are more likely to have dependents: 11 percent of those aged 19–23 have children, while 35 percent of those aged 24–29 and 61 percent of those aged 30–39 have children. Older students are also more likely to be working full-time while taking classes. Overall, students aged 19–23 have an average of 1.2 risk factors; those in the 24–29 age group have 3.2 risk factors, and the 30–39 age group has 3.8 risk factors. Yet these data fail to acknowledge the type of institution—for example, state college, community college, research university, liberal arts college— where students enroll: We do not know which of these students are at community colleges and which are attending other institutions.

Another perspective from NCES data[43] addresses employment for adults, those age 24 and over, who consider work to be their first priority and college to be the second. This group is compared to those who are primarily students but also work as a secondary role. While all of these individuals are considered to be adults by age, their lives are likely to be quite different. Those participants, drawn from the National Postsecondary Student Aid Study, described themselves as employees who also study (56 percent), students who have jobs (26 percent), and students who are not working at the time of the investigation (18 percent). Yet there is no disaggregation in this data set that would show differences between upper socioeconomic status (SES) students at four-year universities and lower SES students at community colleges; thus, several questions remain. Are students who work full-time middle-class adults with mortgages and family commitments, or are they low-wage workers who need their paychecks to pay for groceries and utilities? Are students who are not working at all privileged with financial support from family members, or are they unemployed individuals with few opportunities? While this comparison illustrates differing demographic trends among adult students, it does not describe a complete landscape or an accurate accounting of the adult learner population.

In contrast and focusing upon the significant matter of income status of students, Prince and Jenkins describe adults who work in low-wage jobs and look to community colleges to help them gain any advantage in their economic survival.[44] Using data from the U.S. Department of Education, Prince and Jenkins calculated that "among first-time community college students in 1995–96, 71 percent of those between the ages of 25 and 64 were in the two lower income quartiles, compared with about half of students between 18 and 24" (2). They demonstrate that these low-wage adults make little progress toward a credential, and thus are unlikely to improve their situations economically. Of students aged 25–64 who started in a community college in 1995–96, only 2 percent had earned a bachelor's degree six years later. Four percent had transferred to a university but had no degree, and 11 percent had completed their associate's degree. Seventeen percent finished a certificate, and 6 percent were still enrolled at the community college with no credential attained. The remaining 60 percent were no longer enrolled and were without a credential. Although the economic benefits of attaining various credentials are difficult to verify, the literature (e.g., W. Norton

Grubb)[45] does point to an upward trend in wages for completion of more advanced degrees, as opposed to the student's completion of a small number of credits and subsequent departure from the educational system. Income status of students, particularly for low-wage adult workers, has a bearing upon educational attainment and thus economic benefits for students.

Other terms such as "nontraditional undergraduates" capture a portion of this population, but do not describe it entirely. For example, Choy defines and characterizes "nontraditional undergraduates" as those at any level of postsecondary education: students who delay their entry to college, who carry a part-time academic load, who work while enrolled in college, who are financially independent and may have children or other dependents, who may be single parents, and who do not have high school diplomas. A portion of these students have only one or two of these nontraditional characteristics, while others fit in multiple categories.[46] Choy's data are not disaggregated by age, and thus adult students cannot be separated from the total population. Although there is a substantial body of literature as well as national data sets on the academic progress, enrollment patterns, persistence, and degree attainment of nontraditional students, the connections of this scholarship and the data sets to adult students cannot be verified.

To summarize, data have not been gathered and partitioned in a way to illuminate the condition of disadvantaged adult students. The differences between one scholar's motivated middle-class adults and others' low-wage adult earners can be inferred, but only in part. The use of the same term—adult students—for different populations persists and means that the definition of nontraditional students will be gleaned from contradictory or opaque sources.

Nontraditional Students as a Disadvantaged Population

A useful way to view nontraditional students in higher education institutions in the present day is not as a population characterized by socially constructed traits of age or skin color or physical characteristics (e.g., disability) or by behavioral characteristics or roles connoted by such terms as "dropout," "employee," "parent," "first generation," or "immigrant." Rather, nontraditional students, or what I will call "new" nontraditional students—"new" because we have more sophisticated understanding of this population—are better understood as a disadvantaged population. They are comparatively distinct from a large student population that attends both universities and community colleges. In many respects, the disadvantage can be tied to economic status, but it also includes social, linguistic, and cultural backgrounds and conditions, as well as mental and physical functioning. In some cases, the term *disadvantaged student* could be applied to a majority of students at a single community college. For example, students at East Los Angeles Community College, who are in dire economic conditions, may be a large proportion of students at this college—from 27 percent to 42 percent—and are thus labeled "disadvantaged."[47] The use of the term *disadvantaged* is intended to differentiate this population from the common conceptions of nontraditional students, who may be simply an older population that is over twenty-four years of age.

In this book, I develop this concept of disadvantage and the definition of nontraditional students. I use this concept to ascertain if and the extent to which community colleges as institutions grant justice to this population. That is, I both situate and examine this population within the community college both to understand this population and to judge our educational institutions in their treatment of nontraditional students.

The access mission of community colleges and their accommodation of nontraditional students

The community college is *de facto* an institution for nontraditional students because it serves the most disadvantaged populations in higher education.[48] This institutional characteristic—the condition of its students—has helped to feed the rhetoric of practitioners about the purposes of the community college. George Vaughan compared the community college to America's beacon, the Statue of Liberty, with its appeal to the huddled masses.[49] Indeed, practitioner discourse about the achievements of community colleges approximates religious revival, with an underlying metaphor of saving lost souls. Historian John Frye sees this pattern as practitioners' tendency to view the community college as a path to upward social mobility.[50] Thus, the more disadvantaged the students, the greater the justification for the existence of the community college and its "open access" mission. As well, the small gains of students compared to their characteristics of low-income, poor high school performance, and the like can be exalted as institutional successes. Frye sees another pattern, however, and this is the human resources model, where the labor needs of society are translated into a workforce development role for the community college. Open access is, to some extent, challenged by this model, as external expectations from business and industry are more outcomes-based and standardized, with little attention to individual student progress. During the past two decades, however, the community college has reshaped itself to combine both patterns of individualism connected to personal mobility and social benefits in the form of human resource development through its decidedly economic development mission.[51] This is clearly stated in the proclamation by the national association of community colleges, the American Association of Community Colleges, in 2005: "Community colleges will increasingly be recognized as the gateway to the American dream—the learning resource needed to sustain America's economic viability and productivity."[52] The promise is to sustain open access and at the same time to serve as a nation's principal vehicle for economic development.

A number of forces are pushing more and more low socioeconomic status students to community college. These include demographic shifts, more stringent admissions at four-year schools; program rationalization at four-year colleges; rising college tuition; and workforce training needs and requirements.[53] Thus, while the open access institution is becoming increasingly an instrument of the state and of business and industry, it is also a placement compound for low socioeconomic students, including minority students and immigrants. Again, the question with this condition is whether the institution as it is pressed to serve economic interests can provide justice to disadvantaged populations. Historically, open access has served as a convenient way to signal the accommodation of the disadvantaged.[54]

Open access has variant meanings, including limitations on barriers to participation in the institution and equity of opportunity; but "open" is not an absolute because most programs have selective admissions, from high school completion to test score results and specific course prerequisites, such as Grade 12 mathematics (for technology) or college-level biology (for nursing). On the one hand, the community college's principal mission is arguably access; on the other hand, there are accusations that the community college does not produce outcomes that are consistent with the access principle for beyond college opportunity.[55] This suggests that open access is more about "getting in," absent the promise of postsecondary education—the effects of college on students that vary but nonetheless contribute to both individual and societal development.[56]

I discuss how colleges may provide access to the institution but do not necessarily "accommodate" nontraditional students. I view large numbers of nontraditional students as disadvantaged students by their conditions more than by their specific traits—for example, as members of minority populations. They are disadvantaged because their conditions are not equal to other groups in society—they have not been accorded the same privileges as others whose conditions of birth, wealth, or socially valued assets (such as standardized intelligence) place them in favorable conditions. My view is that institutions must accommodate these disadvantaged students to the extent that their conditions are advantaged or approach equality, even if they cannot be equal in all measures. For higher education institutions, accommodation means those actions and efforts to meet the needs of students—needs that are comprehensive in scope. "Accommodation" is certainly an aspect of access, suggesting that the institution makes room for a wide variety of students. Yet, the term has a customary connotation in student affairs and student persistence literature, suggesting that students will alter or be molded to fit the structures, processes, and norms of higher education institutions.[57] This is evident in the literature on social and academic integration where students assimilate into the norms and patterns of their institution. But accommodation also implies the provision of services, curriculum, and even an environment that satisfies student needs, from tutoring to remedial programs to a campus center for Latino or African American or Native American students. Accommodation can be as straightforward as providing students with health care, as the dean at Mountain View College in Texas explains:

> Health services . . . [is] available to all students. . . . [O]urs is . . . a very small department. We have . . . one limited full-time registered nurse, and then a half-time support staff. . . . [T]heir primary purpose . . . is health

information. . . . [S]ome of our students . . . don't have access to health information. And in some cases they're able to intervene in a way that helps the student maybe get through that day and hang in there. . . . [O]ur special services area is really our program for our students with disabilities. And it's through this area where students identify accommodations, provide the proper documentation. Then we work with the student and the instructor in their classes . . . to try to get the accommodation so they have access to the program.

Accommodation also encompasses educational services that ensure that programs and support are appropriate for the students, as noted by a counselor at Johnston Community College in North Carolina, in a federally funded program for first-generation college students, with at least 33 percent of the population having a learning or physical disability.

Our retention in the program is great. It's much greater than the retention at the college. I actually serve on our quality improvement team for retention, and so we're currently measuring how we retain students as a whole in our college. And in our program, we retained 100% of our students from fall to fall the first year. And this year we anticipate, when you take out our graduates, losing three. And all of those are for financial issues. I think we have [only] three [students] that will not come back in the fall, [so our retention rate] is excellent. I mean, that's just phenomenal.

The entry that we do into our program, the students come in and we say, "How did they find out about [the] program?" That's what we start with. And then we determine if this is really a fit for them. Sometimes they think that we are a funding program. . . . We do not give students money; we give them resources to help them complete. And so once we determine that it's a fit for them, then I talk to them about "what are your goals?" And nine out of ten times it's "I don't know. I don't know. This is what I did for so long and now I want to do something different." And so we immediately do for all of our students, whether they know what they want to do or not, we do a Campbell Interest and Skills Survey. And so using the results of the CISS, that prompts the conversation about what careers are best for them; what courses they should explore. So it's a lot of career exploration and we do goal setting through that. And then, of course, the ones with learning disabilities, I do short-term goal setting, like, "You know you need to set yourself a goal that you're going to get through Math 050 in one semester or that you're going to pass your test next week." For some of them, we meet weekly for goal setting. And then some of them we meet at the beginning of each semester. So, it just depends on their particular needs. (Counselor, Johnston Community College)

Accommodation also suggests, ideally, that students benefit from their college experiences: For example, their education enhances their opportunity for further education or for appropriate employment. If they are going to be dead-ended in minimum-wage jobs because of institutional behaviors that did not advantage them, then their college experience was of little employment benefit. We can conclude that the institution may have failed to accommodate these students.

I equate authentic accommodation with justice—if and the extent to which community colleges provide programs, services, and an institutional climate to advantage students who are disadvantaged. Granted, the institution has multiple influencers, and organizational power is not necessarily in the hands of institutional members. Indeed, over the past two decades, power has arguably shifted from the institution's members—presidents, boards, faculty, and faculty unions—and its community to other influencers, particularly state government and business and industry.[58]

The Study

This book relies upon data from an investigation that began in late 2002 and involves thirteen community colleges in nine states. These sites were selected both purposively and theoretically.[59] They permitted a wide level of access for research and they conformed as a group to criteria that were applied to study nontraditional students in diverse community college settings, including geographical location, institutional size, governance and organizational structures, and demographic characteristics of both the college and its communities. I began by selecting sites containing substantial numbers of minority students or low socioeconomic status students. I reviewed the characteristics of each college's students based upon data from institutional documents. I use data collected at thirteen sites—Pima Community College and GateWay Community College in Arizona, Bakersfield College and El Camino College in California, Community College of Denver in Colorado, Harry Truman College in Illinois, Borough of Manhattan Community College in New York, Johnston Community College and Wake Technical Community College in North Carolina, Mountain View College in Texas, Piedmont Virginia Community College and Virginia Highlands Community College, in Virginia, and Edmonds Community College in Washington. In at least one of these colleges the following characteristics are evident: high percentage of Latino students (Bakersfield, Community College of Denver, Harry Truman College); high percentage of African American students (Wake Technical, Mountain View); large urban college (Borough of Manhattan Community College, El Camino, Pima), rural college (Johnston, Virginia Highlands), and suburban college (Edmonds); multicampus institution (Pima); single campus institution (Edmonds, Johnston, El Camino); and part of a multicollege district (Bakersfield, GateWay, Mountain View, Harry Truman). I also chose a collection of institutions that placed emphasis upon one or more specific curricular orientations: Among these are university transfer (Bakersfield), vocational/technical (GateWay, Wake), developmental (Community College of Denver), remedial (Harry Truman), community and continuing education

(Pima), and comprehensive (Edmonds). I relied upon specific individual administrators at each college who agreed to serve as guides for me and ensure that my access to organizational members, students, and institutional data would be unimpeded.

Interviews were conducted with a sample of administrators, faculty—both full-time and part-time or adjunct—staff, students, and college and state system policy executives. The sample of administrators, faculty, and staff was drawn from criteria that included college members' work responsibilities and their contact with nontraditional students. Those who met the criteria were invited to participate in an interview. The sample of students was selected by student demographics and program areas. Institutional members assisted me with identifying students who represented program areas from remedial to occupational to university transfer, as well as students in continuing and community education courses and programs. As well, students were identified who possessed characteristics of nontraditional students, from minimally nontraditional to "ultra nontraditional." At most of these sites, the college president was either interviewed or engaged in a conversation on the topic of nontraditional students and "students beyond the margins."

Data collection included interviews of selected administrators, faculty, staff, students, and college and state system executives and on-site observations. Approximately 180 interview transcripts were used as sources. Additional sources were institutional documents, such as plans and reports, catalogues, and class schedules, as well as informal conversations and observations. This was a qualitative study that observed the standards of qualitative research data-collection methods, analysis, and design as set out by Baker, Burgess, Le Compte and Preissle, Marshall and Rossman, Mason, Miles and Huberman, and Scott.[60] Data were analyzed according to the guidelines of Miles and Huberman, with pattern coding as a first organizing technique, and several tactics—noting patterns and themes, making metaphors, finding intervening variables, and making conceptual meaning—as techniques for drawing conclusions. In addition, particular attention was given to the narrative line within interviews. Furthermore, data were coded as well through Atlas.ti, a software program,[61] although for this publication the findings were not used extensively.

Data were analyzed using the analytical frameworks of justice[62] to understand institutional practices accorded to nontraditional students, especially "beyond the margins" students, and to identify behaviors of "street-level bureaucrats,"[63] or those college personnel who operate as agents of justice and in some contrast to those structures, such as government policies and economic status of students, that represent forces bearing down on these students. Furthermore, data were analyzed using Mintzberg's power configurations[64] to identify and explain sources and influences of power in community colleges. I use these analytical frameworks to understand and explain conditions and behaviors at community colleges for nontraditional students against a backdrop of a prevailing political economic ideology of neoliberalism. These terms and concepts are addressed in Chapter 2.

In communicating my story of these new nontraditional students and the conflict of justice and neoliberalism, I use considerable dialogue from the conversational interviews that I held with organizational members, students, and state policymakers. These conversations contain narratives, self-expressed and self-reflective accounts of individuals' understandings of themselves and their context. These narratives not only exemplify theories and concepts, but also provide counterpoints and offer a tapestry of the broad spectrum of experiences of nontraditional students.

Chapter 1 begins the discussion of nontraditional students in higher education. The chapter provides a context for understanding the examination of institutions, students, and state systems.

Notes

1. Barbara Ehrenreich, *Nickel and Dimed: On (Not) Getting By in America* (New York: Henry Holt, 2001); Seymour Lipset, *American Exceptionalism: A Double-Edged Sword* (New York: W. W. Norton, 1996); Seymour Martin Lipset, *Continental Divide: The Values and Institutions of the United States and Canada* (New York: Routledge, 1989).
2. Michael Apple, "Comparing Neoliberal Projects and Inequality in Education," *Comparative Education* 37, no. 4 (2001): 409–23; William Bowen and Derek Bok, *The Shape of the River: Long-Term Consequences of Considering Race in College and University Admissions* (Princeton, NJ: Princeton University Press, 1998); Alicia C. Dowd, "From Access to Outcome Equity: Revitalizing the Democratic Mission of the Community College," in "Community Colleges: New Environments, New Directions," ed. Kathleen Shaw and Jerry Jacobs, special issue, *The ANNALS of the American Academy of Political and Social Science* 586 (March 2003): 92–119; Penelope. E. Herideen, *Policy, Pedagogy, and Social Inequality: Community College Student Realities in Postindustrial America* (Westport, CT: Bergin and Garvey, 1998); Sheila Slaughter and Gary Rhoades, *Academic Capitalism and the New Economy: Markets, State, and Higher Education* (Baltimore: Johns Hopkins University Press, 2004); Claude Steele, "Expert Report of Claude M. Steele: Gratz et al. v. Bollinger et al., No. 97-75321 (E.D. Mich.); Grutter et al. v. Bollinger et al., No. 97-75928 (E.D. Mich.)," available at http://www.umich.edu/urel/admissions/legal/expert/steele.html (2003).
3. Robert Reich, *The Work of Nations: Preparing Ourselves for Twenty-First Century Capitalism* (New York: Vintage Books, 1992).
4. Stanley Aronowitz and William Di Fazio, *The Jobless Future: Sci-tech and the Dogma of Work* (Minneapolis: University of Minnesota Press, 1994); Manuel Castells, *The Rise of the Network Society* (Cambridge, MA: Blackwell, 1996); Jeremy Rifkin, *The End of Work: The Decline of the Global Labor Force and the Dawn of the Postmarket Era* (New York: G. P. Putnam's Sons, 1995); Don Tapscott, *The Digital Economy: Promise and Peril in the Age of Networked Intelligence* (New York: McGraw Hill, 1996).
5. Martin Carnoy, *Sustaining the New Economy: Work, Family, and Community in the Information Age* (Cambridge, MA: Harvard University Press, 2000); Richard Florida, *The Rise of the Creative Class: And How It's Transforming Work, Leisure, Community, and Everyday Life* (New York: Basic Books, 2002); Reich, *The Work of Nations*; Rifkin, *The End of Work*.

6. John Levin, *Globalizing the Community College: Strategies for Change in the Twenty-First Century* (New York: Palgrave, 2001); Sheila Slaughter and Larry Leslie, *Academic Capitalism, Politics, Policies, and the Entrepreneurial University* (Baltimore: Johns Hopkins University Press, 1997); Sheila Slaughter and Gary Rhoades, "The Neoliberal University," *New Labor Forum* (Spring/Summer 2000): 73–79.

7. Derek Bok, *Universities in the Marketplace: The Commercialization of Higher Education* (Princeton, NJ: Princeton University Press, 2003); Murray Sperber, *Beer and Circus: How Big-Time Sports Is Crippling Undergraduate Education* (New York: Henry Holt, 2000).

8. Apple, "Comparing Neoliberal Projects"; Bowen and Bok, *The Shape of the River*; David F. Labaree, *How to Succeed in School without Really Learning* (New Haven, CT: Yale University Press, 1997); Slaughter and Rhoades, *Academic Capitalism*; James Valadez, "Cultural Capital and Its Impact on the Aspirations of Nontraditional Community College Students," *Community College Review* 21, no. 3 (1996): 30–44.

9. W. Norton Grubb, "Learning and Earning in the Middle: The Economic Benefits of Sub-Baccalaureate Education" (occasional paper, Community College Research Center, Teachers College, New York, September 1998).

10. Susan P. Choy, "Nontraditional Undergraduates," in *The Condition of Education*, 25–38. U.S. Department of Education, NCES. (2002), Washington, DC: U.S. Government Printing Office. Available at http://nces.ed.gov/pubs2002/2002012.pdf.

11. Choy, "Nontraditional Undergraduates."

12. Kent Phillippe and Madeline Patton, *National Profile of Community Colleges: Trends and Statistics*, 3rd ed. (Washington, DC: Community College Press, American Association of Community Colleges, 2000).

13. Apple, "Comparing Neoliberal Projects"; Noam Chomsky, *Profit over People: Neoliberalism and Global Order* (New York: Seven Stories Press, 1999); Nelly P. Stromquist, *Education in a Globalized World: The Connectivity of Economic Power, Technology, and Knowledge* (Lanham, MD: Rowman and Littlefield, 2002).

14. James Paul Gee, "Identity as an Analytical Lens for Research in Education," *Review of Research in Education* 25 (2001): 99–125.

15. Diana Haleman, "Great Expectations: Single Mothers in Higher Education," *International Journal of Qualitative Studies in Education* 17, no. 6 (2004): 769–84.

16. Ibid.

17. Kathleen Shaw and Sara Rab, "Market Rhetoric versus Reality in Policy and Practice: The Workforce Investment Act and Access to Community College Education and Training," *The ANNALS of the American Academy of Political and Social Sciences* (2003): 172–93.

18. Roueche, John E., and Suanne D. Roueche, *High Stakes, High Performance: Making Remedial Education Work* (Washington, DC: Community College Press, 1999).

19. Christopher Mazzeo, Sara Rab, and Susan Eachus, "Work-First or Work-Study: Welfare Reform, State Policy, and Access to Postsecondary Education," in "Community Colleges: New Environments, New Directions," ed. Kathleen Shaw and Jerry Jacobs, special issue, *The ANNALS of the American Academy of Political and Social Sciences* 586 (March 2003): 144–71.

20. Haleman, "Great Expectations."

21. Gary Teeple, *Globalization and the Rise of Social Reform* (Atlantic Highlands, NJ: Humanities Press, 1995).

22. Apple, "Comparing Neoliberal Projects"; Chomsky, *Profit over People*; Mark Olssen, *The Neoliberal Appropriation of Tertiary Education Policy: Accountability, Research, and Academic Freedom* (2000 [cited May 2004]); available at http://www.surrey.ac .uk/Education/profiles/olssen/neo-2000.htm.
23. John Campbell and Ove Pedersen, "Introduction: The Rise of Neoliberalism and Institutional Analysis," in *The Rise of Neoliberalism and Institutional Analysis*, ed. John Campbell and Ove Pedersen, 2–23 (Princeton, NJ: Princeton University Press, 2001).
24. Simon Marginson and Mark Considine, *The Enterprise University: Power, Governance, and Reinvention in Australia* (New York: Cambridge University Press, 2000).
25. Apple, "Comparing Neoliberal Projects"; Chomsky, *Profit over People*; Stromquist, *Education in a Globalized World*.
26. John Rawls, *Political Liberalism* (New York: Columbia University Press, 1993); John Rawls, *A Theory of Justice* (Cambridge, MA: Belknap Press of Harvard University Press, 1999).
27. Slaughter and Leslie, *Academic Capitalism, Politics, Policies*.
28. John Levin, "The Business Culture of the Community College: Students as Consumers; Students as Commodities," in "Arenas of Entrepreneurship: Where Nonprofit and For-profit Institutions Compete," ed. Brian Pusser, special issue, *New Directions for Higher Education* 129 (2005): 11–26.
29. Levin, *Globalizing the Community College*.
30. Arthur Cohen and Florence Brawer, *The American Community College* (San Francisco: Jossey-Bass, 2003); John Levin, "The Community College as a Baccalaureate Degree Granting Institution," *The Review of Higher Education* 28, no. 1 (2004): 1–22.
31. Gay Bowland, "A Fresh Start," *Community College Week* (August 16, 2004): 6–8; Levin, *Globalizing the Community College*.
32. Sean Creighton and Lisa Hudson, "Participation Trends and Patterns in Adult Education: 1991 to 1999" (Washington, DC: National Center for Education Statistics, 2002).
33. Ali Berker, Laura Horn, and C. Dennis Carroll, "Work First, Study Second: Adult Undergraduates Who Combine Employment and Postsecondary Enrollment" (Washington, DC: National Center for Education Statistics, 2003).
34. National Center for Education Statistics, "Percentage Distribution of Undergraduates, by Age and Their Average and Median Age (as of 12/31/99): 1999–2000," (Washington, DC: U.S. Department of Education, 2000).
35. Teresita L. Chan Kopka, Nancy Borkow Schantz, and Roslyn Abrevaya Korb, "Adult Education in the 1990s: A Report on the 1991 National Household Education Survey" (Washington, DC: National Center for Education Statistics, 1998).
36. Carol Aslanian, "You're Never Too Old: Excerpts from Adult Students Today," *Community College Journal* 71, no. 5 (2001): 56–58. Aslaian's study was done in conjunction with the College Board, and it could be that community college attendees are underrepresented in the data. (N=1,500 undergraduate adults at either level.)
37. Vanessa Smith Morest, *The Role of Community Colleges in State Adult Education Systems: A National Analysis* (New York: Council for Advancement of Adult Literacy, 2004).
38. James Mingle, Bruce Chaloux, and Angela Birkes, "Investing Wisely in Adult Learning Is Key to State Prosperity" (Atlanta: Southern Regional Education Board, 2005).

39. W. Norton Grubb, "Second Chances in Changing Times: The Role of Community Colleges in Advancing Low-Skilled Workers," in *Low-Wage Workers in the New Economy*, ed. Richard Kazis and Marc S. Miller, 283–306 (Washington, DC: Urban Institute Press, 2001).

40. Penelope E. Herideen, *Policy, Pedagogy, and Social Inequality: Community College Student Realities in Postindustrial America* (Westport, CT: Bergin and Garvey, 1998); Kathleen Shaw et al., "Putting Poor People to Work: How the Work-First Ideology Eroded College Access for the Poor" (unpublished document, 2005); Valadez, "Cultural Capital."

41. National Center for Education Statistics, "Percentage of 1999–2000 Undergraduates with Various Risk Characteristics, and the Average Number of Risk Factors" (Washington, DC: U.S. Department of Education, 2000).

42. Choy, "Nontraditional Undergraduates," 25–39.

43. Berker, Horn, and Carroll, "Work First, Study Second."

44. David Prince and Davis Jenkins, "Building Pathways to Success for Low-Skill Adult Students: Lessons for Community College Policy and Practice from a Statewide Longitudinal Tracking Study" (New York: Community College Research Center, Teachers College, Columbia University, 2005).

45. Grubb, "Learning and Earning in the Middle."

46. Choy, "Nontraditional Undergraduates."

47. Research Office, East Los Angeles Community College, "VTEA Student Information Survey, 2001–2002 Accreditation Self-Study," (Los Angeles: East Los Angeles Community College, 2006). This observation was conveyed to me by Estela Bensimon in a personal communication, September 2006.

48. W. Norton Grubb et al., *Workforce, Economic, and Community Development: The Changing Landscape of the Entrepreneurial Community College* (Berkeley: National Center for Research in Vocational Education, University of California, 1997); Herideen, *Policy, Pedagogy, and Social Inequality*; Phillippe and Patton, *National Profile of Community Colleges*; Shaw and Rab, "Market Rhetoric versus Reality"; Jane V. Wellman, *State Policy and Community College-Baccalaureate Transfer* (National Center for Public Policy and Higher Education and the Institute for Higher Education Policy, 2002).

49. George Vaughan, *The Community College Story* (Washington, DC: American Association of Community Colleges, 2000).

50. John Frye, "Educational Paradigms in the Professional Literature of the Community College," in *Higher Education: Handbook of Theory and Research*, ed. John Smart, 181–224 (New York: Agathon Press, 1994).

51. John Levin, Susan Kater, and Richard Wagoner, *Community College Faculty: At Work in the New Economy* (New York: Palgrave, 2006).

52. American Association of Community Colleges, "Our Mission Statement" (American Association of Community Colleges, January 2005), available at http://www.aacc.nche.edu/Content/NavigationMenu/AboutAACC/Mission/OurMission Statement.htm.

53. Richard Alfred and Patricia Carter, "New Colleges for a New Century: Organizational Change and Development in Community Colleges," in *Higher Education: Handbook of Theory and Research*, ed. John C. Smart and William G. Tierney, 240–83 (New York: Agathon Press, 1999); Thomas R. Bailey and Irina E. Averianova, "Multiple Missions of Community Colleges: Conflicting or Complementary" (occasional paper, Community College Research Center, Teachers

College, New York: 1998); Steven Brint, "Few Remaining Dreams: Community Colleges since 1985," in "Community Colleges: New Environments, New Directions," ed. Kathleen Shaw and Jerry Jacobs, special issue, *The ANNALS of the American Academy of Political and Social Sciences* 586 (March 2003): 16–37; Cohen and Brawer, *The American Community College*; Levin, Kater, and Wagoner, *Community College Faculty*; Slaughter and Rhoades, *Academic Capitalism*.

54. Richard Richardson and Louis Bender, *Fostering Minority Access and Achievement in Higher Education* (San Francisco: Jossey-Bass, 1987); Richard Richardson, Elizabeth Fisk, and Morris Okun, *Literacy in the Open-Access College* (San Francisco: Jossey-Bass, 1983); John E. Roueche and Suanne D. Roueche, *Between a Rock and a Hard Place: The At-Risk Student in the Open-Door College* (Washington, DC: American Association of Community Colleges, 1993).

55. Quentin Bogart, "The Community College Mission," in *A Handbook on the Community College in America*, ed. George Baker, 60–73 (Westport, CT: Greenwood, 1994); Brint, "Few Remaining Dreams"; Steven Brint and Jerome Karabel, *The Diverted Dream: Community Colleges and the Promise of Educational Opportunity in America, 1900–1985* (New York: Oxford University Press, 1989); John Dennison and Paul Gallagher, *Canada's Community Colleges* (Vancouver: University of British Columbia Press, 1986); W. Norton Grubb, *Honored but Invisible: An Inside Look at Teaching in Community Colleges* (New York: Routledge, 1999).

56. Ernest T. Pascarella and Patrick Terenzini, *How College Affects Students*, (San Francisco: Jossey Bass, 1991); Ernest T. Pascarella and Patrick Terenzini, *How College Affects Students: A Third Decade of Research* (San Francisco: Jossey Bass, 2005).

57. Dudley Woodard, "Comments on Research" (personal communication, Raleigh, NC, 2005).

58. Levin, *Globalizing the Community College*; John Levin, "Public Policy, Community Colleges, and the Path to Globalization," *Higher Education* 42, no. 2 (2001): 237–62; Barbara Townsend and Susan Twombly, eds., *Community Colleges: Policy in the Future Context* (Westport, CT: Ablex, 2001).

59. Margaret Le Compte and Judith Preissle, *Ethnography and Qualitative Design in Educational Research* (San Diego, CA: Academic Press, 1993); Catherine Marshall and Gretchen Rossman, *Designing Qualitative Research*, 3rd ed. (Thousand Oaks, CA: Sage, 1999); Jennifer Mason, *Qualitative Researching* (Thousand Oaks, CA: Sage, 1996).

60. Carolyn Baker, "Ethnomethodological Analyses of Interview," in *Handbook of Interview Research: Context and Method*, ed. Jaber Gubrium and James Holstein, 777–95 (Thousand Oaks, CA: Sage, 2002); Robert Burgess, *In the Field: An Introduction to Field Research* (London: George Allen and Unwin, 1984); Robert Burgess, *Strategies of Educational Research: Qualitative Methods* (London: Falmer Press, 1985); Matthew Miles and A. Michael Huberman, *Qualitative Data Analysis* (Thousand Oaks, CA: Sage, 1994); J. Scott, *A Matter of Record* (Cambridge: Polity Press, 1990).

61. Atlas.ti Ver. Win 5.0 (Build 63), qualitative analysis software from ATLAS.ti GmbH (formerly Scientific Software Development), Berlin.

62. Rawls, *A Theory of Justice*.

63. Martin Lipsky, *Street-Level Bureaucracy* (New York: Russell Sage Foundation, 1980).

64. Henry Mintzberg, *Mintzberg on Management: Inside Our Strange World of Organizations* (New York: Free Press, 1989).

Chapter 1

Nontraditional Students
in Higher Education

This chapter presents frameworks for the understanding of nontraditional students in higher education, reviews the scholarly literature, discusses national data on nontraditional students, and identifies themes drawn from both the data and the scholarly literature. It sets the stage for the argument that follows in Chapter 2 and offers evidence for the significance of nontraditional students in higher education as well as their condition as students.

Theoretical orientations and assumptions that underlie inquiry not only frame what researchers, scholars, and interpreters see, but also explain the phenomena from a particular ontological perspective. The concepts and language we use to examine behaviors serve as metaphors or tropes that both enlarge and limit our understandings.[1] How we frame, identify, and understand nontraditional students reflects our ontological preferences and shapes how we see the behaviors or experiences of these students. I identify and name three broad frameworks that are employed in the scholarly literature and by practitioners to view and discuss nontraditional students. They include the trait framework, the behavioral framework, and the action framework. Each carries with it not only particular points of view about these students, but also a set or system of evidence that both supports claims and perpetuates the framework itself. Most important, each framework defines the problem of or for nontraditional students—for example, "at-risk" for noncompletion of college programs,[2] isolation from other student groups,[3] and unjust institutional treatment.[4]

Frameworks for Understanding the Issues

At least three dominant frameworks guide our understanding of nontraditional students and their conditions in higher education. First is the trait framework, which suggests a deficit model for understanding and indeed responding to students. This deficit model is centered upon the institution and how the institution can fulfill its purposes and goals and operate with the students it enrolls. This construction views students as possessing traits that hinder or help them in their college endeavors and that pressure or shape the institution. From this construction, ascribed identity concepts such as "at risk," "first generation," "mature," "minority," and "underprepared" are attached to students, as well as institutional strategies to respond to these characteristics. These strategies are found in various practices such as "learning communities," "assessment of student learning outcomes," "first-generation programs," and curricular and instructional initiatives and interventions, such as outcomes-based curriculum, mentoring programs, and service learning.[5] Because the trait framework encompasses the largest body of literature and is the dominant one in higher education, this chapter devotes considerably more discussion to it than the others.

Second is the behavioral framework, which incorporates how students experience and view college. This framework draws upon students as actors, a phenomenological perspective that highlights the issues from the student point of view, including student motivations and goals as well as their learning and developmental outcomes. This framework is exemplified in the work of Kathleen Shaw, Robert Rhoads, and James Valadez, among others.[6]

Third is the action framework, which addresses how the institution, the state, and indeed the public and private sectors treat and behave toward the student. I include within this framework the policies and actions of institutions and the state, such as welfare reform, that affect students.[7] This framework incorporates the justice model that I derive from John Rawls.[8] The action framework indicates if and the extent to which the community college and its mission and actions provide justice for students.

In my discussion of each framework, I use both the scholarly literature and my collected data from interviews to describe and explain the frameworks. The three frameworks and their associated assumptions are displayed in Table 1.1.

The Trait Framework

Research and policy studies that address nontraditional students in higher education in general use the trait framework and rely largely upon conceptions of individual traits to identify and define nontraditional students: age, gender, social and economic status, educational attainment, attendance patterns, employment status, familial relationships, ethnic or racial identity, citizen and immigrant status, geographical location (rural, urban, suburban), and the like. Using the trait conception, Susan Choy estimates that approximately 75 percent of postsecondary education students are nontraditional.[9]

Frameworks for Understanding Nontraditional Students

Given this figure, traditional does not equate with a minor population of students, but rather, we can speculate, with an historical or customary population that attended or attends college. Traditional students continue to be viewed as the norm. Furthermore, the public emphasis upon this traditional student population and the maintenance of a trait- based conception of students have

Table 1.1 Frameworks for Understanding Non-Traditional Students

	Trait framework	*Behavioral framework*	*Action Framework*
Focus	Characteristics of students	Behaviors of students	Institutional and government policies and behaviors
Assumption	Students missing or deficient in specific qualities	Students exhibit resiliency or nonresiliency, motivational levels, and goals	Actions and policies of institutions and governments shape student behaviors and influence student outcomes
Purposes	Student persistence in college	Student experience of college	Treatment of students

assisted practitioners, researchers, and scholars to focus upon all students from what I refer to as an institutional perspective—a view that frames these students within the political economy of higher education. From the institutional perspective, nontraditional students with identifiable traits are differentiated from traditional students, and in this comparison nontraditional students are seen as deficient—in academic background, in economic status, in possessing social and cultural capital—and thus less likely to meet the standards, expectations, and markers of attainment than traditional students who are not deficient. Thus, the term "at risk" students is attached to nontraditional students because they possess characteristics or traits that impede their progression in college at the same rate or at the same level as the norm—that is, traditional students. The literature is replete with unequal outcomes for nontraditional students who are defined using a trait perspective.[10] These unequal outcomes expand beyond the usually cited outcomes of degree attainment and program completion and range from the digital divide—lack of computer skills, lack of proficiency in using the Internet, and lack of access to computer[11]—to earning potential based upon institutional attendance or program of study.[12]

Characteristics of Nontraditional Students within the Trait Framework

As distinct from 1970, when 25 percent of postsecondary students were nontraditional, by 1999, 73 percent of postsecondary students were classified as nontraditional. By the 2000s, the figure rises to 75 percent: Nontraditional students are now the rule, not the exception.[13] These students actually comprise the majority of community college students in credit programs. In some sectors, nontraditional students are defined as those over the age of twenty-four who are engaged in some form of postsecondary learning activities—a definition synonymous with "adult learners."[14] In all higher education institutions, 37 percent of undergraduate students are twenty-five or older; in four-year institutions this percentage is 22.8 percent of the total undergraduate population, and in two-year colleges the percentage is 36.8 percent of all students.[15] However, there are other characteristics used to categorize these students, such as their marital status and whether they have children. This population of adult learners is often the "invisible" portion at many four-year institutions and makes up a large part of the students at community colleges.[16]

From the perspective of characteristics, students with at least one characteristic of nontraditionality, as defined by the U.S. Department of Educational Research, comprise 90 percent of all community college students.[17] That is, nontraditional students are numerically synonymous with the total population of community college students. Thus, it is not uncommon for community college practitioners to either express disdain for the label "nontraditional," resist its use, or consider all their students to be identical—whether labeled traditional or nontraditional.

> A traditional student is an adult; the traditional student is not an eighteen-year-old kid who shows up at a university and lives in a dorm. That is not America. So when you are talking about nontraditional students, I think that higher education and particularly researchers and others . . . start out with a premise that is not accurate; it is wrong; it is not well-rounded. So I don't know how far down the path that takes you if you begin with the idea that the traditional student is the eighteen-year-old kid who lives in a dormitory. That was 1940s, to be quite honest with you. So for us, the traditional student is an adult [who has] got a family. . . . In my perception [it] is the eighteen-year-old [who] is the nontraditional. (Chancellor, Pima College)
>
> [F]rankly every student at BMCC is a nontraditional student. There's no longer the traditional typical student who's going to come to school during the day. . . . We offer courses from seven in the morning to ten at night, from Sunday to Saturday, on-site, off-site, online, because we're addressing the needs of all the students and all their issues. and it involves many areas, not just academically but . . . support services and kinds of support services that we offer the students for free. (Dean of Academic Programs, Borough of Manhattan Community College)
>
> [We] don't discriminate . . . here as traditional or nontraditional. I have at other colleges talked about traditional, nontraditional. . . . If you pressed me I'd say a traditional student would be the typical student that either comes right out of high school or a few years out of high school and gets into a transfer or vocational program, finishes, goes to work or transfers. . . . [T]he nontraditional student has become our traditional student. . . . [A]ll of our students have some measure of nontraditional background or behavior or preparation or something of that nature. So I think we've just become a place for students to start their educational career in whatever way they want to do it. (President, Edmonds Community College)

Nonetheless, even those practitioners who do not accept the traditional/non-traditional distinctions do categorize students, whether by age—"the only distinction I really make is day and evening student, where our day students tend to be younger and our evening students tend to be older" (President, Borough of Manhattan Community College)—or by academic and social background—"we can put that in a broad category of saying the students are underprepared socially too, because it's not only the academic part of it where we're saying they have to take remedial" (Associate Dean, Borough of Manhattan Community College). Others who accept the concept of "nontraditional" have contextual understandings and applications of the term. The president of Johnston Community College in North Carolina views nontraditional students as those who are distinct populations within the majority of community college students.

[Nontraditional students] would be community college students with special needs: not your traditional definition of nontraditional, but rather your specialized populations within that adult population. [They are] first . . . generation college students, economically and academically disadvantaged and disabled, the reentry mother or father for now; especially now [there are] more fathers coming back because they've been downsized. [Nontraditional is] just the specialized; just the few specialized populations. (President, Johnston Community College)

Several specific characteristics of these nontraditional students—their gender, race or ethnicity, and socioeconomic status—are key to their identification as a discrete group in postsecondary education. According to data for the 1999–2000 academic year from the National Center for Education Statistics (NCES),[18] nontraditional students in all of higher education are more likely to be women, black, and in the lowest income group. The combination of characteristics or traits leads to statistical assertions about attendance and attainment. Women from lower socioeconomic status were more likely to enroll in a community college than their middle- and upper-class counterparts. These women were also found to be older than their male peers: The average age of male nontraditional students is twenty-six, and the average for females is twenty-seven. The average age for white students is twenty-six; blacks, twenty-seven; Latino, twenty-six; Asian, twenty-five; American Indian/Alaskan Native, twenty-eight; Native Hawaiian/Pacific Islander, twenty-six; biracial/multiracial, twenty-six; and other, twenty-five.[19] Black students are among the oldest in age, with American Indians reported as the oldest students on average. Whites and Latinos are the same age, on average, at twenty-six.

NCES data from the 1999–2000 year give percentages of all postsecondary students classified as nontraditional (where the definition is based upon age, that is, twenty-four years of age or older) by race or ethnicity.[20] Forty percent of white students are nontraditional. Among other racial or ethnic groups, 48 percent of black students are nontraditional; 42 percent of Latinos; 41 percent of Asians; 53 percent of American Indian/Alaska Natives; 42 percent of Native

Hawaiians/Pacific Islanders; and 37 percent includes other and more than one race. Clearly, higher proportions of minority students than white students are nontraditional. With an increase in minority student enrollment and the growth in the number for nontraditional students, there is an expansion of nontraditional students of various races and ethnicities in higher education.

There are a disproportionate number of underrepresented or minority students, compared to white students, with low socioeconomic backgrounds attending institutions of higher education. Socioeconomic status (SES) of students in postsecondary education indicates that minority students—particularly blacks, Latinos, American Indians, and Hawaiian/Pacific Islanders—occupy the lowest rung. Nineteen percent of white students are low SES, as compared with 37 percent black; 35 percent Latino; 32 percent Asian; 25 percent American Indian; 31 percent Hawaiian/Pacific Islander; and about 30 percent of other or more than one race.[21] Minority status combined with socioeconomic status may indeed qualify a student as nontraditional. Such a conception informs research on student persistence that compares these populations to the norm—the traditional student.[22]

Access and Educational Outcomes for Nontraditional Students within the Trait Framework

How do these differences regarding student characteristics play out? Recent research on community colleges by Thomas Bailey et al. indicates that institutional effects—that is, what institutions contribute to students—have little influence upon student outcomes for degree completion or retention at college compared to student effects—that is, individual student characteristics.[23] Thus access alone may not lead to outcomes that are customarily valued—degree attainment—or used as a measure of attainment. Nontraditional minority students enter higher education later in life than their white peers due in part to the disparity in secondary education and the access to social capital that these students receive.[24] Watson Scott Swail et al. refer to this condition as the "educational pipeline" that minority students pass through to reach postsecondary education.[25] They found that Latino students do not have the same access to social capital as their white peers. For example, 35 percent of white students had parents with at least a bachelor's degree, whereas only 14 percent of Latino students possessed a similar educational legacy. For Latino students, higher education is less likely to be an achievement aspiration. While 78.3 percent of white students aspired to receive at least a bachelor's degree, only 55.2 percent of Latino students shared that goal. This explains in part why minority students wait until later in their lives to pursue higher education: As secondary education students poised to continue on to postsecondary education, they are not influenced by their families and peers to pursue higher education. Their white peers, however, do receive the types of social capital messages at home that encourage them to participate in higher education. Latino students and students of other minority groups often wait until higher education is necessary to acquire the credentials and skills needed for a better job—a vocational aspiration—or to show their own children the importance of education—a familial relationship value.

Nontraditional students as a whole are less likely than traditional students to persist to degree completion or to remain enrolled in a community or four-year college. A research project from 2002 that focused upon student attainment classified 9 percent of surveyed students as high-risk candidates for program completion and 66 percent as moderate risk; that is, a total of 75 percent of all 33,500 students in their pool were classified as being "at risk" of failing or dropping out of college.[26] More specifically, students who are poor or who have backgrounds that classify them as low socioeconomic status are at greater risk of either not completing postsecondary education programs or graduating. Among this group, young people and minority students have the lowest educational outcomes.[27] There is no doubt that students' characteristics—age, race, socioeconomic status, gender, geographical location, dependency status, and the like—are linked as multivariables of student academic attainment.

The trait framework is useful in identifying those populations whose educational attainment is either more precarious or challenging, or both, than either a majority or traditional population or even a comparative population, such as females compared to males. Although women have made up the majority of community college students since the late 1970s,[28] their majority status has not always translated into superiority in attainment or in educational conditions. In 2001, women comprised 57.3 percent of the total national enrollment.[29] These gains in enrollment also translated into gains in degree attainment. Katharin Peter and Laura Horn state that "by 2001, women of all racial/ethnic groups [excluding nonresident aliens] earned a majority of the degrees awarded. In particular, black women earned two-thirds of both associate's degrees and bachelor's degrees awarded to black students. Hispanic and American Indian women were awarded 60 percent or more of associate's and bachelor's degrees conferred to Hispanic and American Indian undergraduates, while Asian women earned 57 percent of associate's degrees and 55 percent of bachelor's degrees conferred to Asian students."[30]

While women earn the majority of associate's and bachelor's degrees, other markers of attainment are not as commendable. Peter and Horn find that "women make up 60 percent of students in the lowest 25 percent income level, 62 percent of students age 40 or older, 62 percent of students with children or dependents (among married or separated students), and 69 percent of single parents. All of these characteristics are associated with lower rates of persistence and completion in postsecondary education."[31] Women are less likely than men to complete the transfer from a community college to a bachelor's granting college, even though more women attend community college than men. Additionally, those women who do transfer are less likely to earn a bachelor's degree than are men.[32] The impact of both transfer behavior and baccalaureate attainment on females is evident in their wage earning and career advancement: Women receive less monetary benefit from their educational endeavors than men. Female associate's degree earners can expect to earn an average salary of $28,324, while males with the same credential can expect an average of $44,655. Even women with bachelor's degrees earn less ($38,045.55) than men with associate's degrees.[33]

Women of color often face more barriers and have a lower level of economic benefits from education than white women. In 1999, the median annual earnings for women working full-time was $30,900 for white women, $27,600 for African American women, $25,500 for Native American women, and $23,200 for Latinas. Asian American women actually had the highest average, at $33,100.[34]

While the reasons that women give for attending college do not differ significantly from those of men, reasons given by women for discontinuing attendance include health concerns, family situations, and emotional or financial difficulties.[35] In contrast, men tend to report dropping out because coursework is not challenging. The majority of women attend the community college part-time (63.9 percent of all female community college students in fall 2001) and have other responsibilities, including family and work,[36] which no doubt affect academic progress and degree completion.

Of undergraduate students in 1999–2000, 43 percent were twenty-four years old or more. Fifty-six percent of that age group defined themselves primarily as employees who also study. Twenty-six percent of that age group said they were primarily students who also had to work, and only 18 percent were not employed at all. The average age of those who considered themselves to be employees was thirty-six, while the average for the students who work was thirty. Women constituted 56 percent of both groups; that is, for people twenty-four or older who are combining work and study in some proportion, they are more likely to be female. This percentage rises when looking at students age 30–39 (61 percent female) and students age 40 or older (63 percent female).[37]

Another population identifiable through the trait framework that occupies a more precarious position for postsecondary education attainment is that group classified as disabled. There are several different types of disabilities that can impact a person's educational path, according to data from the National Longitudinal Transition Study-2 (NLTS2). Examples of disabilities include speech impairments, orthopedic impairments, autism, mental retardation, emotional disturbances, traumatic brain injuries, learning disabilities, visual or auditory impairments, other health impairments, or multiple disabilities.[38] In general, the persistence rate of students with disabilities through high school is fairly good. High schools are regulated by IDEA (Individuals with Disabilities Education Act), which requires schools to take all necessary steps to allow the student to complete the curriculum, including individualized education plans.[39] For students in the NLTS study with visual or hearing impairments, the rates of graduating from high school are high (90 percent to 95 percent). Students with orthopedic impairments or autism also have good success rates (86 percent to 88 percent). The students with the lowest reported rate of high school completion are those with emotional disturbances (56 percent), multiple disabilities (65 percent), mental retardation (72 percent), or learning disabilities (75 percent).

Postsecondary entry and persistence, however, are more uneven among youth with disabilities. Even those who received sound academic preparation from their high schools may struggle with the transition to college. Higher education is governed by the ADA (Americans with Disabilities Act) rather than IDEA, and therefore has a different legal definition of accommodation.[40] Only 31 percent of youth with disabilities take postsecondary classes after leaving high school. Community colleges are the preferred destination for most of these students. Twenty percent of youth with disabilities take classes from community colleges; 6 percent participate in postsecondary vocational, business, or technical schools; and 9 percent attend initially, or eventually, four-year colleges.[41] Lynn Newman indicates that only 35 percent of postsecondary students with documented disabilities receive services or accommodations, whether from lack of making the request or lack of qualifying for services at the college level.

Student characteristics or traits have multiple and differing influences upon student participation in college and levels of attainment. In the case of disabled students, motivations for attending postsecondary education and expectations for performance have salience. Students who envisioned themselves as participating in postsecondary coursework were more likely to do so, at a rate of 36 percent versus 5 percent of students who did not have that goal.[42] From the trait perspective, people with disabilities are the least likely population to attend postsecondary education. Within that population, there is a wide gap in postsecondary participation between those whose self-ascribed identity includes college attendance and those whose identity does not include attendance. However, the scholarship on disabled students in postsecondary education is limited, since it conceives of college attendance as participation in degree programs.

Given the wide variation in characteristics of nontraditional students from the point of view of the trait framework, including the degree of nontraditionality identified by Choy and Horn and Carroll,[43] there is both a practical and scholarly need to differentiate among the various categories of nontraditional students. Figure 1.1 displays this differentiation as a pyramid structure. Not all nontraditional students are the same or have the same "at risk" conditions. For example, those who are classified as minimally nontraditional may simply be students who have postponed higher education for one year.

Indeed, within the trait framework, scholarship that addresses nontraditional students has been aimed primarily at degree-enrolled students. At community colleges, this category excludes those students who are enrolled in noncredit courses or engaged in coursework even though they do not anticipate applying to a degree program. Thus, studies of at-risk students in the community college literature fail to acknowledge these students who may constitute a large portion of community college students. I refer to a substantial portion of these students as "beyond the margins," or "ultra nontraditional students," to distinguish them from mainstream credit students or from students who are identified with "marginalized" groups, particularly those with minority status, but who enroll in credit programs. Higher education scholars largely have ignored this population, with some exceptions, and they do not fit in the pyramidal structure identified by scholars such as Choy or Horn and Carroll (see Figure 1.1).[44]

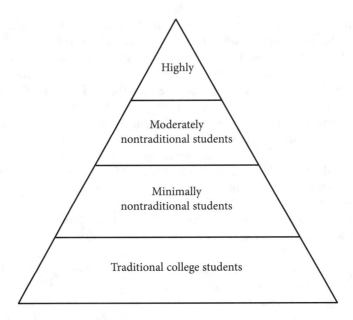

Figure 1.1. Traditional and nontraditional students

Minimally nontraditional students exhibit one characteristic of nontraditionality, such as identity as an underrepresented minority. Moderately nontraditional students exhibit two or three characteristics, such as not only identity as an underrepresented minority, but also as a reentry student and as someone in need of financial aid because of a low economic status. A highly nontraditional student exhibits a multitude of characteristics; the student not only has minority and reentry status and financial need, but also works over twenty hours a week and may be a single parent. Those beyond the customary framework of nontraditional students not only possess characteristics of nontraditional students, but they may also engage in programs that include noncredit continuing education (which encompass both noncertificate and externally certified programs and courses), contract training provided for employers, and credit continuing education programs that are not part of the mainstream of delivery. These are students at least on the periphery of education if not outside the periphery so as to be invisible.

"Beyond the Margins" Students

Welfare recipients, the working poor, students with both physical and mental disabilities, and undocumented immigrants who are non-English-speaking constitute a portion of that class of students who are either ignored in scholarship or merged with students who have distinctly different characteristics and face less harrowing circumstances. In incorporating a new class of students into the concept of nontraditional students, I see a continuum of nontraditionality, building upon the work of Horn and Carroll and Choy, from students who are minimally nontraditional to students who are beyond the customary nontraditional framework.[45] Figure 1.2 represents this group in comparison to traditional and nontraditional college students.

Recent work on welfare and education—by Jacobs and Winslow; Mazzeo, Rab, and Eachus; London; and Shaw, Goldrick-Rab, Mazzeo, and Jacobs—has alerted us to a different population of students who access the community college: welfare mothers.[46] Welfare mothers not only qualify for nontraditional student status, but also for status as students beyond the margins. They are found in remedial courses, externally certified programs of six months or less, and in programs designed for paid return to work. These students are rarely captured in enrollment data sets.

In addition to welfare recipients who attend the community college, there are other disadvantaged populations who are students. They include the working poor, or "low-wage workers,"[47] as well as those whose jobs have vanished. Tereaza, who came from Yugoslavia in 1985, attends Truman College in Chicago to complete a high school equivalency program (GED). To support herself and her family, she undertakes a number of low-paying jobs: "Jobs, I have a lot of odd jobs, like cleaning jobs," she says. Al at Virginia Highlands Community College is unemployed, formerly a company worker—a technician and warehouseman—who is attending college with the help of state grants for dislocated workers. "I'm unemployed. The company I worked for shut down after fifteen years," he says. "I started school before it shut down, just taking a class at a time, getting ready for this. And soon, as they shut it down, I started full-time in the fall."

Adult immigrants are another group who may be considered "beyond the margins." Those who are seeking higher education are likely to turn to the community colleges first, due to lower tuition rates, accessible schedules and locations, and availability of relevant programs such as English as a second language (ESL). Government websites indicate that the majority of immigrants currently entering the United States are coming from Mexico, India, the Philippines,

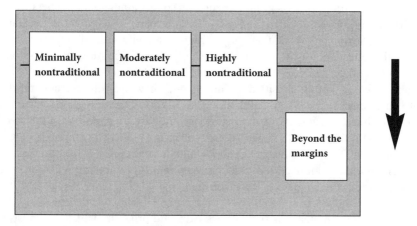

Figure 1.2. Beyond-the-margins students

mainland China, Vietnam, and the Dominican Republic. Urban centers such as New York, Los Angeles, Chicago, San Francisco, and Miami have traditionally been destinations for immigrants, but newer arrivals are entering rural areas as well. Carl Bankston[48] indicates that 26.8 percent of all foreign-born individuals residing in the United States were living in the South by the year 2000. These recent immigrants are likely to have limited English skills and lower incomes, both of which are relevant to educational participation. Community colleges have responded to these changes: In 1991, only 40 percent of community colleges provided ESL classes, but that increased to 55 percent in 1999.

This group or class of nontraditional students—"beyond the margins" students—are often segregated from other nontraditional students. This segregation can be physical, in the form of discrete programs for this population only, and often programs are held in facilities that are detached from mainstream college programs or from the campus altogether. Segregation can also occur in program legitimacy—whether this means a low status program or a noncredit program or an underfunded program, such as one where state appropriations are significantly less that those for other college programs. Furthermore, segregation can have a linguistic basis, as in the case of programs for ESL students, where these students have limited understanding of the students in their institution and little connection between these students and other college students and services. In this sense, because there are few if any institutional effects, student characteristics dominate and certainly are predictive of student outcomes. The question, of course, is whether the institution is doing enough to advantage these students so that their disadvantages are not the critical variable in their college experiences or outcomes.

While mainstream community college students might conform to the human capital model of the community college,[49] suggesting that students are potential workers and thus economic investments, those students who are outside the mainstream, outside the margins, are almost invisible not only to scholars, policymakers, and government officials, but also to administrators and faculty in their own institution. This group of students counts among its members many who are in noncredit courses and programs, who are or have been on welfare, who suffer from long-term unemployment, who are physically disabled or mentally challenged, and who are often people of color. These students are beyond the bureaucracy: They are rarely captured in national or even state data-collection machinery; often classified as "noncredit," they also can be students who are physically separate from the mainstream in off-campus programs—at work sites, in church basements, or in prisons. Yet, they are a component of the institution that claims to be an "open access" college.

Earl, a GED full-time instructor and faculty union official at Truman College in Chicago, offers an overview of a large segment this student population, one which dominates the focus of his institution.

At Truman, because it is such a huge Adult Ed program, which is almost all ESL, we are the largest single source of students going into the credit program. I think . . . 30% of the students going into credit programs come from us. That is because Truman has a huge ESL [English as a second language] program. I think we have 11,000 students and over 10,000 of them are immigrants with . . . over hundred languages. . . . We are one of the three legs, really: adult continuing education, our [ESL] students who are noncredit [and] receive state and federal funding, and then the credit students that pay tuition.

According to Earl, at Truman College, adult education students are "people [who] drop out of high school or they never finish, [and] they drop out for a million different reasons. Usually it has to do with kids that are poor; their friends are into drugs; they want to go out and play; there is alcoholism or drug problems in the family; they have to go to work; they are pregnant: you know, all of those kind of things, plus a million different other stories."

Grace, an ESL instructor at Johnston Community College in North Carolina, identifies the characteristics of students who are both nonnative speakers and socially disadvantaged.

I'm helping people learn English; I'm helping them . . . just become, be able to become, part of our society, be able to interact and do things that we take for granted: Go shopping, open a bank account, obtain a library card. If you don't have the language, that's a very difficult thing to do. . . . [T]he majority are Hispanic, although I have had Filipinos, Iranians, and a couple of other nationalities, but it's mainly Hispanic and mainly Mexicans and Hondurans. They're all over eighteen, of course; it's an adult program. Some of my students have been studying loyally and regularly for a year and the progress is just incredible. They can now order a pizza by phone. We set goals in the class. . . . Now they have library cards; we're learning to use the Internet. I mean, their worlds are expanding. (Grace, ESL Instructor, Johnston Community College)

In discussing federal welfare programs, Lourdes, the dean of Adult Basic Education at Wake Technical Community College in Raleigh, North Carolina, not only confirms the research conclusions of Kathleen Shaw, Sara Goldrick-Rab, Christopher Mazzeo, and Jerry Jacobs,[50] but also characterizes a student population ignored in national data-gathering efforts, making these students almost invisible.

[With] all the welfare reform programs . . . it will only get worse. It's "get to work quick." Not, "let's train you in a trade so that you are going to be able to support your family. . . . We want to get you off the road and we want to get them out as quickly as possible." They want a quick fix, but they find out they still cannot support their family. Then what are they working for? There is a group who has limited social services benefits. . . . After they get a job . . . they can stay there for a year. After that year is up, they are going to have to quit work, because they can't afford to work and pay for someone to take care of their kids. They don't make enough to work and keep their kids in day care. . . . We have the young mothers, both black and white. You have as many stories as [to] why they left high school [and are] dropouts. (Dean of Adult Basic Education, Wake Technical Community College)

The trait framework, then, groups and categorizes students by their characteristics and by either their deficiencies— economic, learning, physical, and the like—or their statistically ordained future—such as "at risk" for degree completion—or both characteristics and deficiencies. With respect to "justice as fairness,"[51] some categories of students are more disadvantaged than others: Single-parent African American women are more disadvantaged than twenty-year-old single white males, all other traits (such as physical and mental ability) being equal. Although not aligned with justice but rather addressing student performance and attainment, the trait framework suggests that unless there is institutional intervention, the educational results of the more disadvantaged will not only be lower than those of the less disadvantaged, but also below the standards of acceptability for all students.

From the trait perspective alone, the role of the institution is framed as an interventionary agent, meditating between student characteristics and institutional factors as well as external influences both on students and on the institution. These influences include federal and state policy, such as student financial aid and targeted programs, the economy and labor market, and other educational institutions and employers. The capacity and ability of institutions to intervene may be limited with respect to effects upon student attainment, given that the scholarly evidence suggests that institutional factors in persistence and educational attainment, while statistically significant, are small.[52]

The Behavioral Framework

Student as agent, as an actor who has both experiences and perceptions, is the key concept of this framework. From this perspective we view and understand individual students and groups of students as possessing identities that both shape and influence experience and perception. Kenneth Gonzalez investigates not only how students with Chicano identities experience the university environment, but also how these students "negotiate" their experiences in the academic process. Furthermore, he offers self-reflection in recalling his own student experience in the context of his Chicano identity.[53] The identity as student is but one facet of the multiple identities of those who attend college. Liz, at Edmonds Community College in Washington State, is not only a student but also an employee of the college, whose life at fifty-two years of age has been both complicated and a struggle. Yet, she also looks beyond her own difficulties and reaches out to help others.

There's been downtimes, but there's been more when I look around and I see some of the students that don't have food or something, because I have four students that work for me at night. I know a few of them are living on their own and barely making it, so I'm always bringing in soup or finding stuff at the store, and I say, "OK, guys, ration it out for the week because we're not keeping it," or else I'll say, "You know, honey, I need some money because I'm going to go out and buy the kids some food, some real food. Not some pizza or something." (Liz, Edmonds Community College)

Dennis, at Community College of Denver, has a complicated and painful set of identities; he is not only a returning adult student but also a cancer patient. He is working his way through courses at the same time he is battling against a life-threatening disease.

My name is Dennis. . . . I'm forty-one years old; I'm married for the second time with a two-and-a-half-year-old son. First time, my first marriage was to a woman for fifteen years. That fell apart because of her drug use. I worked for the government for fifteen years. Four children with her; they're all grown. . . . I did asbestos removal. I worked with plutonium, uranium, zinc, lead. I was a superintendent and a project planner designer. I was hurt on a project out at Rocky Flats in 1992. I lost two of my men. I went in to get the rest of the men out and I lost two. . . . I'm out of the field; I became disabled. I haven't done nothing, really, for the last ten years, except try to get my life back together from the hospitals and all that stuff, and deal with cancer and all that. Beat them battles. So, I'm on top of that. (Dennis, student, Community College of Denver)

One student I encountered in a class for students with physical and cognitive challenges at Bakersfield College was a younger student, twenty years old, who had appeared to have a number of either cognitive or physical disabilities, or both. When he told me his story, it became quite clear what this individual was facing. He lived in a family of fully deaf people. There were five brothers and sisters and two parents. They were all deaf. He was the only hearing person in his family. I asked him what he wanted to do and why he was at the college. He indicated to me that he was there to learn how to be an interpreter for deaf people, and that he had practice because he was the only one in his family who was not deaf. He could sign, but he wanted to perfect his skills. I noted to him that he had some trouble with speech and some difficulty with concentration. I asked him a little more about his background and he indicated to me that when he was six years old, he had had a stroke, a rare occurrence, and that that stroke partially debilitated him, made learning extremely difficult for him. To this day, he is not able to read. I looked at him and I said, "How are you able to learn so well when you cannot read?" And he said, "I have very good hearing. I have a very good memory." He had recently completed high school and was attending Bakersfield College.

Within this framework, nontraditional students become students with disadvantages who struggle, because of their identities, against economic, institutional, and social constraints.[54] If viewed through a trait framework, these individual students would be labeled "at risk," possessing one or more risk characteristics that suggest their chances of course and program completion are hampered because they lack particular attributes. The larger picture, however, is that these students struggle to reach personal and educational goals. Some of them are impeded by barriers, such as family responsibilities, lack of resources, discrimination, or their own lack of interest, lack of information, or lack of guidance. Others not only endure through personal hardships but indeed overcome obstacles, demonstrating resilience.[55] These are students such as Dishon who is blind and Alicia who is deaf. For Dishon, lack of sight means dependency on his dog, Franklin, and significant effort to work in order to pay for his college studies in business management. For Alicia, learning English as her second language—American Sign Language is her first—has meant considerable struggle in writing and reading so that she can progress in her college-level studies. Others—single mothers who have little financial support, dislocated workers, and those who are disabled—exhibit a high level of resilience, transcending the categories that predict their lives through the criteria of traits. Dennis from Community College of Denver exemplifies not only the trials but also a high level of resilience.

> Having some more medical issues coming up, because of the battles I beat and all the stuff. . . . I'm a go-getter. I'm not going to quit. But as far as schooling, I talked about it with [my wife], discussed it with her for about a year and half, and the fear was to come back and, I guess, look like a dummy for not knowing all the stuff. I came in and I tested, and the testing that I did do that I thought I would be higher on, I was the lowest on. So it worked opposite on me. I don't know why that happened, but it did, and one of the subjects is math. I'm in 030 in math. . . . I've been out of math for . . . years and you just can't remember. . . . I've got all As in my classes except one. One class . . . I'll probably get a C out of it. I'm learning how to format and do essays and all that stuff, and I came from an F in that class all the way up to a D, to like a C. . . . Now [the instructor] says, "You'll walk out of here with a C or a high C or . . . a B." She says, "You're doing all right." She's real proud of me. (Dennis, student, Community College of Denver)

The behavioral framework provides us with another definition of nontraditional students, one that is not based on traits, such as age or race/ethnicity. We understand students and the condition of their nontraditionality by looking at their experiences in their pursuit of personal and educational goals through their attendance at a postsecondary institution.

The Action Framework

This framework focuses upon the actions of others—individuals, groups, organizations, institutions, and government, as well as business and industry—as directed to nontraditional students. In the main, it addresses how nontraditional students, particularly disadvantaged students, are treated by institutions. A handful of scholars characterize the community college as comprised of students who are from marginalized groups or are oppressed as a consequence of their life experiences, or both.[56] The efforts of these scholars point us to understanding the variety of community college students and the larger scope and implications of access. Valadez[57] examined what he refers to as "nontraditional students," in the early 1990s, who are in the 2000s the majority of community college students, noting that the institution did not allocate sufficient resources or remove barriers for student achievement. These actions prevented students from moving into academic program tracks that are the pathway to social and economic advancement. Herideen acknowledges the specific plight of returning women, who are viewed as people with collapsed lives and lacking in self-esteem, to flourish in the community college, and she offers emancipation *à la* Freire,[58] seeking to raise the consciousness of these vocational students. While these are students previously referred to as "at-risk students,"[59] the focus of researchers—namely, Herideen, Valadez, and Rhoads; Shaw, Rhoads, and Valadez; and Valadez—is not aimed at improving institutional performance in student retention or achievement measures. Rather, this work addresses institutional treatment—accommodation—of students, particularly students who are disadvantaged.

Coeli is from Chihuahua, Mexico, and has lived in Denver for four years, where she graduated from high school with a 4.0 grade point average. She is not a U.S. resident, but she has aspirations to complete a master's degree in architecture. Her college grades in all subjects are high, and she is enrolled in Calculus. From high school, she was accepted to the University of Colorado, but she could not afford the out-of-state tuition, which she would have to pay as an undocumented immigrant. She was able to secure scholarships from the Community College of Denver through their first-generation program.

> I didn't go to CU [University of Colorado] because I'm not a U.S. citizen, and they charge $32,000 a year. . . . I got accepted and everything, but they just consider me international and there's no way to pay that. And I also got a scholarship. . . . They pay for everything, everything, computer, everything . . . but the law, the rules changed . . . three months before I got the scholarship and . . . so I didn't [go to university because I was not a resident]. I wanted to go to a university, not community college, first of all. And I was, like, really disappointed when I was coming here because it was just a community college and I wasn't going to have the college experience. . . . And since I joined this program [first generation], I got more involved. . . . I used to get, like, really sad because I thought I wasn't going to go to college because, like, everyone told me, "Well, if you're not a U.S. citizen you can't go to college." But now that I'm going here and I know that I'll do whatever to keep studying, even go to another state or whatever. (Coeli, Arts and Sciences university transfer student, Community College of Denver)

Coeli is disadvantaged by the state as a nonresident—state law requires inequitable treatment for her if she attends the University of Colorado. However, the Community College of Denver provides her with scholarships so that she can attend that institution without paying tuition or fees, or for books and supplies. In this sense, the Community College of Denver compensates for her inequitable treatment by the university system and treats her with justice. Nonetheless, Coeli will again be disadvantaged once she completes her associate's degree and transfers to a public university, where her residency status will require her to pay out-of-state tuition, or to a private university that can grant her degrees in architecture or architectural engineering and not provide her with scholarships.[60]

Of considerable significance to the construction of this framework are ideological underpinnings of the community college, particularly those pertaining to students. The multipurpose nature and the multiple missions of the community college, amply detailed in literature on the community college,[61] may obfuscate the central mythos of the institution that has been amplified and illuminated in the work of numerous scholars, including those who are boosters of the institution and those who are critics. This mythos or plot[62] takes the form of underserved student populations—what George Vaughan alludes to as the "huddled masses"[63]—or individual students with disadvantages who may or may not transcend their personal, social, economic, or cultural limitations. The positive path of this narrative, one more commonly found in the practitioner literature, is that students with disadvantages arrive at community colleges with seemingly insurmountable limitations or barriers facing them. Through a combination of their motivation and the institution's efforts, these disadvantaged students fulfill or exceed their potentials in educational attainment.[64] The negative path of the narrative, one more commonly found in the scholarly literature, is that the institution does not live up to its promises as a vehicle for social mobility and democratizing agent. These disadvantaged students, then, either underperform or simply perform at the level expected for their station and condition.[65] While the tension is between dreams fulfilled or dreams diverted, the mythos focuses upon the institution's role in student attainment. One side of this tension is expressed by a dean at Community College of Denver.

> [E]very student [is] one who's moving on. . . . [T]hat might be university, but that doesn't mean that I know that they will [make it], but I like to think of that. . . . I truly believe that people rise to the standards that you set and rise to your expectations. . . . I look at students as they're going somewhere and we're helping them move forward, and one of the ways to do that is to understand that wherever they go, they need to be able to succeed there. . . . [W]hat I try to do when I look at students is how high are you aiming and how do we get you to aim higher.

The other side is explained by two state-level community college officials in Colorado.

> [Y]ou look at all this budget pressures, if you are . . . [a] college president and you want to succeed, then why would you direct your efforts to an at-risk student, when we've got waves of students coming who are

academically prepared, who have demand for the classes, and who can succeed? We invest in them; we can move them on; we can do a good job; we can produce some degree under these conditions. If we focus our efforts on that at-risk student, then we do so at risk ourselves.

I think in Colorado because of the budget cuts . . . all of the social aspects of what we do have been sacrificed. . . . [I]f an at-risk student comes in . . . they're not treated any differently [from other students]. . . . [If they] don't succeed, well, we're sorry. . . . [W]e did our best. But nobody has the resources or the inclination . . . to say we're going to treat you special because you came to us for help and we're going to help you. I think maybe we could've done that in the past.

That is, the view of idealism and optimism for students is counterpoised against the views of system and institutional incapacity to fulfill these ideals, in this case because of limited resources and lack of political will.

Thus, the action framework for nontraditional students guides us in the examination of the treatment of students by institutions and their officials, by government, as well as by other entities, including both the public and private sectors. This framework will identify both institutional and public policies that either thwart or enhance student access to and attainment in postsecondary education. As in the example of Coeli, one segment of public higher education discriminates against her and frustrates her aspirations and achievements, while another segment offers her equitable treatment and thus accommodates her fully into the institution. This framework will illuminate our understanding of Rawls' theory of justice, specifically justice as fairness and the principle of fair equality of opportunity,[66] which is addressed in Chapter 2.

Notes

1. Gareth Morgan, *Images of Organization* (Thousand Oaks, CA: Sage, 1997).
2. John E. Roueche and Suanne D. Roueche, *Between a Rock and a Hard Place: The At-Risk Student in the Open-Door College* (Washington, DC: American Association of Community Colleges, 1993).
3. Kenneth Gonzalez, "Campus Culture and the Experience of Chicano Students in Predominantly White Colleges and Universities" (paper presented at the annual meeting of the Association for the Study of Higher Education, San Antonio, TX, November 18–21, 1999).
4. John Levin, "Nontraditional Students and Community Colleges: The Conflict of Justice and Neoliberalism. An Overview" (paper presented at the American Educational Research Association, Montreal, April 2005).
5. John E. Roueche, Eileen E. Ely, and Suanne D. Roueche, *In Pursuit of Excellence: The Community College of Denver* (Washington, DC: Community College Press, 2001).
6. Kathleen M. Shaw, "Defining the Self: Constructions of Identity in Community College Students," in *Community Colleges as Cultural Texts*, ed. Kathleen M. Shaw, J. Valadez, and R. Rhoads, 153–71 (Buffalo: State University of New York Press, 1999); Kathleen Shaw, Robert Rhoads, and James Valadez, eds., *Community Colleges*

as Cultural Texts (Albany: State University of New York Press, 1999); James Valadez, "Cultural Capital and Its Impact on the Aspirations of Nontraditional Community College Students," *Community College Review* 21, no. 3 (1996): 30–44.

7. Penelope E. Herideen, *Policy, Pedagogy, and Social Inequality: Community College Student Realities in Postindustrial America* (Westport, CT: Bergin and Garvey, 1998); Rebecca A. London, "The Role of Postsecondary Education in Welfare Recipients' Paths to Self-Sufficiency" (Santa Cruz: University of California–Santa Cruz, 2004); Kathleen Shaw and Sara Rab, "Market Rhetoric Versus Reality in Policy and Practice: The Workforce Investment Act and Access to Community College Education and Training," *The ANNALS* (2003): 172–93.

8. John Rawls, *A Theory of Justice* (Cambridge, MA: Belknap Press of Harvard University Press, 1999).

9. Susan P. Choy, "Nontraditional Undergraduates," in *The Condition of Education*, 25–38. U.S. Department of Education, NCES (2002), Washington, DC: U.S. Government Printing Office, available at http://nces.ed.gov/pubs2002/2002012.pdf.

10. Betty A. Allen, "The Student in Higher Education: Nontraditional Student Retention," *Community Services CATALYST* 23, no. 3 (1993): 19–22; Alexander Astin and Leticia Oseguera, "The Declining 'Equity' of American Higher Education," *The Review of Higher Education* 27, no. 3 (2004): 321–41; Thomas R. Bailey, Mariana Alfonso, and Marc Scott, "The Education Outcomes of Occupational Postsecondary Students" (New York: Community College Research Center, Teachers College, Columbia University, 2005); Jason De Sousa, "Reexamining the Educational Pipeline for African-American Students," in *Retaining African Americans in Higher Education: Challenging Paradigms for Retaining Students, Faculty, and Administrators*, ed. Lee Jones, 21–44 (Sterling, VA: Stylus, 2001); Susan E. Kent and Michael J. Gimmestad, "Adult Undergraduate Student Persistence and Their Perceptions of How They Matter to the Institution" (paper presented at the Association for the Study of Higher Education, Kansas City, MO, 2004), 1–21; Gregory S. Kienzl, "The Triple Helix of Education and Earnings: The Effect of Schooling, Work, and Pathways on the Economic Outcomes of Community College Students" (paper presented at the Association for the Study of Higher Education, Kansas City, MO, 2004), 1–62; Christine A. Ogren, "Rethinking the 'Nontraditional' Student from a Historical Perspective: State Normal Schools in the Late Nineteenth and Early Twentieth Centuries," *The Journal of Higher Education* 74, no. 6 (2003): 640–64; Ernest T. Pascarella and Patrick Terenzini, *How College Affects Students*, (San Francisco, CA: Jossey-Bass, 1991); Derek V. Price, "Defining the Gaps: Access and Success at America's Community Colleges," in *Keeping America's Promise*, 35–37 (Indianapolis, IN: Lumina Foundation, 2004); John Roueche and Suanne Roueche, *Between a Rock and a Hard Place*; Minerva Santos, "The Motivations of First-Semester Hispanic Two-Year College Students," *Community College Review* 32, no. 3 (2004): 18–34.

11. Kent A. Phillippe and Michael J. Valiga, *Faces of the Future: A Portrait of America's Community College Students* (Washington, DC: American Association of Community Colleges, 2000), 1–8.

12. Pascarella and Terenzini, *How College Affects Students*.

13. Choy, "Nontraditional Undergraduates."

14. Brian Pusser and Andrea Spreter, "A Framework for a Meta-analysis of Research on Adult Learners" (Charlottesville: University of Virginia, 2005); Richard Voorhees and P. Lingenfelter, "Adult Learning and State Policy" (Chicago: State Higher Education Executive Officers and Council for Adult and Experiential Learning, 2003).

15. "College Enrollment by Age of Students, Fall 2003," in *The Chronicle of Higher Education: Almanac Issue 2005–2006* (Washington, DC: Chronicle of Higher Education, 2005).

16. Carol E. Kasworm, "Adult Student Identity in an Intergenerational Community College Classroom," *Adult Education Quarterly* 56, no. 1 (2005): 3–20.

17. Choy, "Nontraditional Undergraduates."

18. NCES, "The Condition of Education 2000" (Washington, DC: U.S. Department of Education, 2000); NCES, "The Condition of Education 2001" (Washington, DC: U.S. Department of Education, 2001); NCES, "The Condition of Education 2002" (Washington, DC: U.S. Department of Education, 2002); NCES, "The Condition of Education 2003" (Washington, DC: U.S. Department of Education, 2003); NCES, "The Condition of Education 2004" (Washington, DC: U.S. Department of Education, 2004).

19. NCES, "The Condition of Education 2002."

20. Ibid.

21. Ibid.

22. Bailey et al., "The Effects of Institutional Factors on the Success of Community College Students" (New York: Community College Research Center, Teachers College, Columbia University, 2005); Pascarella et al., "First-Generation College Students: Additional Evidence on College Experiences and Outcomes," *The Journal of Higher Education* 75, no. 3 (2004): 249–84; Pascarella and Terenzini, *How College Affects Students: A Third Decade of Research* (San Francisco: Jossey-Bass, 2005).

23. Bailey et al., "The Effects of Institutional Factors."

24. Kenneth Gonzalez, Carla Stoner, and Jennifer Jovel, "Examining Opportunities for Latinas in Higher Education: Toward a College Opportunity Framework" (paper presented at the annual meeting of the Association for the Study of Higher Education, Richmond, VA, 2001): 2–37.

25. Watson Scott Swail et al., "Latino Students and the Educational Pipeline" (Virginia Beach, VA: Educational Policy Institute, 2005).

26. Community College Survey of Student Engagement, "Engaging Community Colleges: A First Look" (Austin, TX: Community College Leadership Program, 2002). This group of surveyed students constitutes mainstream students. That is, they are degree-seeking students, not students in noncredit or certificate programs. This latter group I later refer to as "students beyond the margins," many of whom are ignored in survey research.

27. London, "The Role of Postsecondary Education in Welfare Recipients' Paths to Self-Sufficiency"; Roueche, Ely, and Roueche, *In Pursuit of Excellence*; Roueche and Roueche, *High Stakes, High Performance: Making Remedial Education Work* (Washington, DC: Community College Press, 1999).

28. ERIC Development Team, "Two-Year College Students: A Statistical Profile" (Los Angeles: ERIC Clearinghouse for Junior Colleges, 1982).

29. Kent Phillippe and Leila Gonzalez Sullivan, *National Profile of Community Colleges: Trends and Statistics*, 4th ed. (Washington, DC: American Association of Community Colleges, 2005).

30. Katharin Peter and Laura Horn, "Gender Differences in Participation and Completion of Undergraduate Education and How They Have Changed over Time" (Washington, DC: U.S. Department of Education, National Center for Education Statistics, U.S. Government Printing Office, 2005).

31. Ibid., v.

32. Brian Surette, "Transfer From Two-Year to Four-Year College: An Analysis of Gender Differences," *Economics of Education Review* 20, no. 2 (2001): 151–63.

33. Phillippe and Gonzalez Sullivan, *National Profile of Community Colleges.*

34. Amy Caiazza, April Shaw, and Misha Werschkul, "Women's Economic Status in the States: Wide Disparities by Race, Ethnicity, and Region" (Washington, DC: Institute for Women's Policy Research, 2004).

35. Jennifer Wolgemuth, Nathalie Kees, and Lynn Safarik, "A Critique of Research on Women Published in the *Community College Journal of Research and Practice*: 1990–2000," *Community College Journal of Research and Practice* 27, nos. 9–10 (2003): 757–68.

36. Phillippe and Gonzalez Sullivan, *National Profile of Community Colleges.*

37. Ali Berker, Laura Horn, and C. Dennis Carroll, "Work First, Study Second: Adult Undergraduates Who Combine Employment and Postsecondary Enrollment," (Washington, DC: National Center for Education Statistics, 2003).

38. Mary Wagner, "Characteristics of Out-of-School Youth with Disabilities," in *After High School: A First Look at the Postschool Experiences of Youth with Disabilities*. A report from the National Longitudinal Transition Study-2 (NLTS2), ed. Mary Wagner et al. (Menlo Park, CA: SRI International, 2005).

39. Committee on Education and the Workforce, Individuals with Disabilities Education Act (IDEA): A Guide to Frequently Asked Questions (cited December 2005), available from http://edworkforce.house.gov/issues/109th/education/ideaidea faq.pdf.

40. Lynn Newman, "Postsecondary Education Participation of Youth with Disabilities," in Wagner, ed., *After High School*.

41. Ibid.

42. Ibid.

43. Choy, "Nontraditional Undergraduates"; Laura J. Horn and C. Dennis Carroll, "Nontraditional Undergraduates: Trends in Enrollment from 1986–92 and Persistence and Attainment among 1989–90 Beginning Postsecondary Students," NCES 97-578 (Washington, DC: U.S. Department of Education, Office of Educational Research and Improvement, 1996).

44. Choy, "Nontraditional Undergraduates"; Horn and Carroll, "Nontraditional Undergraduates."

45. Ibid.

46. Jerry A. Jacobs and Sarah Winslow, "Welfare Reform and Enrollment in Postsecondary Education," in "Community Colleges: New Environments, New Directions," ed. Kathleen Shaw and Jerry Jacobs, special issue, *The ANNALS of the American Academy of Political and Social Sciences* 586 (March 2003): 194–217; London, "The Role of Postsecondary Education"; Christopher Mazzeo, Sara Rab, and Susan Eachus, "Work-First or Work-Study: Welfare Reform, State Policy, and Access to Postsecondary Education," *The ANNALS of the American Academy of Political and Social Sciences* (March 2003): 144–71; Kathleen Shaw et al., "Putting Poor People to Work: How the Work-First Ideology Eroded College Access for the Poor," unpublished manuscript (2005).

47. Lisa Matus-Grossman and Susan Tinsley Gooden, "Opening Doors to Earning Credentials: Impressions of Community College Access and Retention from Low-Wage Workers," (Paper presented at the annual research conference of the Association for Public Policy Analysis and Management, November, Washington, DC: 2001).

48. Carl Bankston, "Immigrants in the New South: An Introduction," *Sociological Spectrum* 23, no. 2 (2003): 123–28.
49. Herideen, *Policy, Pedagogy, and Social Inequality*.
50. Shaw et al., *Putting Poor People to Work*.
51. Rawls, *A Theory of Justice*.
52. Pascarella and Terenzini, *How College Affects Students: A Third Decade of Research*, vol. 2 (San Francisco: Jossey-Bass, 2005).
53. Kenneth Gonzalez, "Inquiry as a Process of Learning About the Other and the Self," *Qualitative Studies in Education* 14, no. 4 (2001): 543–62.
54. Marion Bowl, *Nontraditional Entrants to Higher Education* (Stoke on Trent, UK: Trentham Books, 2003).
55. I am grateful to Richard Newman at the University of California, Riverside, for this observation in a discussion of my research. A colleague of mine at North Carolina State University, Colleen Wiessner, also offered a similar observation by pointing me to a work on resilience: Gina O'Connell Higgins, *Resilient Adults: Overcoming a Cruel Past* (San Francisco: Jossey-Bass, 1994).
56. Herideen, *Policy, Pedagogy, and Social Inequality*; Robert Rhoads and James Valadez, *Democracy, Multiculturalism, and the Community College* (New York: Garland, 1996); Kathleen Shaw, Robert Rhoads, and James Valadez, "Community Colleges as Cultural Texts: A Conceptual Overview," in *Community Colleges as Cultural Texts*, ed. Shaw, Valadez, and Rhoads (Albany: State University of New York Press, 1999), 1–13.
57. Valadez, "Cultural Capital."
58. Paulo Freire, *Pedagogy of the Oppressed* (New York: Continuum, 1994).
59. Roueche and Roueche, *Between a Rock and a Hard Place*.
60. Whereas Coeli could attend Community College of Denver as an undocumented immigrant paying resident fees, by 2005 Higher Education policy in Colorado altered so that undocumented immigrants at all public postsecondary institutions were classified as non-residents and compelled to pay non-resident fees.
61. Bailey and Irina E. Averianova, "Multiple Missions of Community Colleges: Conflicting or Complementary" (occasional paper, Community College Research Center, Teachers College, New York, 1998); Bailey and Vanessa Smith Morest, "The Organizational Efficiency of Multiple Missions for Community Colleges" (New York: Teachers College, Columbia University, 2004); Arthur Cohen and Florence Brawer, *The American Community College* (San Francisco: Jossey-Bass, 2003); John Frye, "Educational Paradigms in the Professional Literature of the Community College," in *Higher Education: Handbook of Theory and Research*, ed. John Smart, 181–224 (New York: Agathon, 1994); W. Norton Grubb, *Honored but Invisible: An Inside Look at Teaching in Community Colleges* (New York: Routledge, 1999); David F. Labaree, "From Comprehensive High School to Community College: Politics, Markets, and the Evolution of Educational Opportunity," *Research in Sociology of Education and Socialization* 9 (1990): 203–40; John Levin, "Missions and Structures: Bringing Clarity to Perceptions About Globalization and Higher Education in Canada," *Higher Education* 37, no. 4 (1999): 377–99; John Levin and J. Dennison, "Responsiveness and Renewal in Canada's Community Colleges: A Study of Organizations," *The Canadian Journal of Higher Education* 19, no. 2 (1989): 41–57; John Roueche and George A. Baker III, *Access and Excellence* (Washington, DC: Community College Press, 1987).
62. Northrop Frye, *Anatomy of Criticism* (Toronto, ON: McClelland and Stewart, 1966).

63. George Vaughan, *The Community College Story* (Washington, DC: American Association of Community Colleges, 2000).

64. Roueche and Baker, *Access and Excellence*; Roueche, Ely, and Roueche, *In Pursuit of Excellence*.

65. Steven Brint and Jerome Karabel, *The Diverted Dream: Community Colleges and the Promise of Educational Opportunity in America, 1900–1985* (New York: Oxford University Press, 1989); Kevin Dougherty, *The Contradictory College* (Albany: State University of New York Press, 1994); Rhoads and Valadez, *Democracy, Multiculturalism, and the Community College*.

66. Rawls, *A Theory of Justice*.

Chapter 2

Theoretical Frameworks

Over the past thirty years, the community college has become not only an educational institution that invites two somewhat contradictory views of its performance, but also the home of increasing numbers of students who do not possess characteristics of students at universities or traditional baccalaureate four-year colleges. There is considerable debate about the purposes and accomplishments of the community college, and much of the debate centers around its role in moving students toward social and economic equality with others in higher education.[1] This is no doubt a question of the definition of justice. For several scholars, the academic and employment outcomes of community college students must be equal to those of students in four-year colleges and universities for community colleges to be viewed as worthy institutions that grant justice to their students.[2] Although the practitioner literature does not make such comparisons, it nonetheless makes claims that community college students are well served by the institution and that outcomes, particularly employment, are advantaging students.[3]

Judgments about the community college, based upon student outcomes, become increasingly testy and significant in a higher educational context where there are changing population demographics, considerable growth in student numbers, and more overt competition for students, prestige, and resources. For example, in 1990, there were 13.8 million higher education students; in 2002, that number had grown to 16.6 million.[4] During that same period, community college enrollments rose from 5.2 million to over 6.5 million. The minority student population of community colleges was 33.3 percent, up from 26.5 percent just a decade earlier. During this period, approximately 64 percent of students attended community colleges part-time, compared to approximately 20 percent at four-year colleges.[5] These figures, combined with the academic background and economic conditions of community college students, suggest that the community college has considerably more work to do with its students than four-year colleges and universities in order to equalize both outcomes and opportunity for students compared with those institutions. Unless institutional effects have more influence upon student performance than the literature indicates,[6] the community college cannot hope to equalize outcomes for its students compared to those at four-year colleges and universities. Thus, if justice is defined as creating conditions or outcomes of equality, then the community college falls short.

Related to the demographic pattern, the quest for prestige or greater legitimacy that characterizes the present trend in higher education[7] not only separates
the community college from other higher education institutions, such as the liberal arts college or the research university, but it also affects the mission of the
community college, as that institution, too, is not immune from advancing its status. The community college baccalaureate degree is a case in point.[8] Ultimately,
community college students must compete with university students, and clearly
the community college's low status role impedes the chances of its students.[9]

The political and economic context for both community colleges and other
higher education institutions engages these institutions in competition over
both students and resources.[10] In order to compete, the community college has,
to a large extent, followed the business model of operations, choosing efficiency
as one method to satisfy its sponsors and to gain and satisfy customers as an
open access institution that endeavors to meet the demands of the public.[11] The
use of part-time faculty—now making up 67 percent of the total faculty workforce—and "one stop" centers for students are two important examples.[12] Within
a global economy, the community college has modeled both services and production upon corporations.[13] With its increasing emphasis upon workforce
development and its default role as a remedial center to prop up those who are
in need of upgrading for employment or preparation for further academic education,[14] the community college has become a vehicle for neoliberal policies.[15]

This chapter provides an explanation of the theories of justice, neoliberalism,
and globalization that serve as foundations for the perspectives I use in discussing the book's topics. I rely upon a body of scholarship in both higher education and the sociology of work to explain neoliberalism as well as the views of
neoliberal critics such as Noam Chomsky.[16] Furthermore, I explain understandings of neoliberalism from the perspective of higher education. This chapter
also introduces and explains the concepts of organizational power and "street-
level bureaucrats" and shows their application to higher education institutions.
New managerialism is also introduced as another way to view the effects of
neoliberalism upon the academy. Finally, the chapter concludes with a discussion of the apparent and theoretical conflict between neoliberalism and justice.

In examining institutional behaviors and their actors, I rely upon several concepts and theoretical perspectives. The first and principal concept, justice, is
based upon John Rawls' theory of justice—justice as fairness and the principle of
fair equality of opportunity. For Rawls (1999), the hardships of some are not ruled
out by the greater good to the aggregate or whole: "[I]n order to treat all persons
equally, to provide genuine equality of opportunity, society must give more attention to those with fewer native assets and to those born into the less favorable
social positions."[17] In these community college sites, I determine the extent to
which those in less favorable social and economic stations are accorded justice.

Specifically, I use Rawls' "difference principle," which states that "social and
economic inequalities . . . are to be adjusted so that, whatever the level of those
inequalities, whether great or small, they are to be the greatest benefit of the
least advantaged members of society."[18] As well, I adopt his argument of a social

contract between members of a society, a contract that implies a future, applicable from one generation to the next. This permits me to look at not only individuals and their fair treatment, but also groups and classes of people to determine fairness over time. As well, I apply Rawls' notion of good, or "rational advantage." That is to say that advantage in a cooperative arrangement applies to individuals, groups, and institutions. In my use, I apply this good to students, to college members, and to the college itself, as well as to society. Finally, I adopt Rawls'view of good as a developing element, and this affords me the opportunity to speak to both evolving conditions at community colleges as well as potential conditions that I can propose.

Rawls opposes a utilitarian conception of justice, wherein maximizing collective happiness is the preeminent goal. Instead, Rawls argues that unequal distributions of wealth and power cannot be justified merely by aggrandizing cumulative utility: "All social values . . . liberty and opportunity, income and wealth, and the bases of self-respect . . . are to be distributed equally unless an unequal distribution of any, or all, of these values is to everyone's benefit."[19] This, I suggest, speaks to the issue of privilege in higher education, wherein merit is equated with benefits, justifying elite educational opportunities for a small percentage of those who have demonstrated academic achievement in high school and whose continued advantage will arguably benefit the public good.[20]

For my analyses, I incorporate Rawls'second principle and its corollaries into my interpretation of data. According to Rawls, each person is to have an equal right to the most extensive basic liberty, compatible with a similar liberty for others, and social and economic inequalities are to be arranged so that they are both (a) reasonably expected to be to everyone's advantage and (b) attached to positions and offices open to all. This view is consistent with underlying principles of the community college mission, which includes an "access for all" imperative and an espoused goal of democratization.[21]

In applying Rawls to educational institutions, I conclude that institutions must be judged by how effectively they guarantee the conditions necessary for all equally to further their aims, and by how efficiently they advance shared ends that will similarly benefit everyone. In other words, we can judge a nation's or a state's educational apparatus by how well it facilitates actual, not merely formal, equal opportunity for the worst-off citizen. We can consider developmentally challenged students, for example, as among the worst off in society, as they lack those basic skills required for independent living. According to Rawls, educational institutions have a responsibility to ensure substantive equality of opportunity, regardless of the potential economic benefits of unequal access. Specifically, disadvantaged students must not be subjected to an educational system or program in which their individual agency and self-purpose are neglected in favor of the economic benefit for a local industry.

Theoretically, I rely upon neoliberalism and globalization theory[22] and their application to higher education institutions, particularly the economic influences upon organizational behaviors. While proponents of neoliberalism argue that individuals are rewarded based upon their achievement, critics of neoliberalism deride this view and indicate that those who have favorable conditions—those with privilege—are unjustly rewarded, and that rewards are based more on capabilities than achievement.[23] This argument certainly has implications for and applications to higher education institutions, where there are contests over issues of merit and inequality.[24] My previous work on globalization and the community college touches on some of the fallout from a competitive economy: that education becomes increasingly oriented toward higher-level programming, credentials, and economic marketplace demands.[25] This orientation results in a replay of economic globalization within the institution, where there are "winners" and "losers" and some programs are favored and other programs are neglected and sometimes jettisoned. That work did not take a close look at those programs and students on the losing side. In this research, I examine programs and students less favored in an economically competitive environment, those that have low academic capital.[26]

I use neoliberalism, in particular, to examine the concept of lifelong learning and to critique that concept. This is amplified in Chapter 7. In this critique, I rely upon a body of scholarship in both higher education and the sociology of work.

Overall, I view higher education institutions as value systems. At the core of recent musings on the behaviors of higher education institutions is the questioning of values in higher education.[27] Research scholarship takes a less moralistic tone but nonetheless is value-laden in its examinations. It is evident from these two bodies of literature that higher education institutions are reflections of value systems, containing, on the one hand, a number of value systems and serving as vehicles for value systems outside the institution. More specifically, although the values embedded in the community college mission may suggest some contrariness and tensions,[28] that mission historically connotes public accessibility to education and training and "open door" admissions, including the admission of those normally left out of other postsecondary institutions. From this perspective, I consider the condition of the access mission of the community college and the accommodation of those at the lower end of the economic spectrum. To this extent, my work connects with institutional theory, particularly neoinstitutionalism.[29]

Furthermore, I examine the community college as a human service organization,[30] to determine its focus upon students outside the mainstream; to note the "fallout" from a competitive economy; and to identify the practice of justice within an institution. I rely upon a considerable body of community college literature that characterizes the community college, at least in part, as an institution whose mission includes attention to the least favored segments of society,[31] and as an institution that addresses the social and economic needs of its community, especially at basic levels of individuals who are coping with daily life. To this end, I draw upon theories of justice as well as principles of communitarianism that extol the benefits of social interest and value individual development and responsibility.[32] In addition, I find that Mintzberg's power theory,[33] particularly his concept of power configurations, is useful. These configurations are conceptualizations—ideal typologies, according to Mintzberg—that are the amalgam of internal and external organizational conditions and influencers. Organizational actions, according to Mintzberg, are determined by arrangements or configurations of power—from those with considerable control exerted by external sources or influencers, such as governments, to those with a high level of control of an executive group or autocratic leader within an organization. As well, I rely upon the concept of "street-level bureaucrats"[34] to explain the actions of college administrators, faculty, and staff, as well as policy executives at both the institutional and state level. Michael Lipsky notes that public service organizations, with vague or ambiguous goals and the tendency to ration services, permit considerable discretion among managers and staff to make personal judgments in their daily work. He refers to these individuals who exercise personal judgment in their dealings with customers or clients as "street-level bureaucrats."

I use these perspectives to analyze the perceptions of college members on students and programs and the behaviors of community colleges in their treatment of students, particularly "beyond the margins" students. For example, Mintzberg's missionary configuration is especially appropriate for those institutions or units within institutions that hold strong belief systems and apply these to the students. Lipsky's "street-level bureaucrat" concept helps me to understand how institutions accommodate students beyond the margins: For example, faculty and administrators do not ask and do not tell which of their students are illegal immigrants, and thus these students are provided English language training on the same basis as legal immigrants, residents, and citizens.

As I am also interested in the state's role in promoting social and economic justice, I rely upon Carnoy[35] to analyze the state's use of the community college as an extension of capital and corporate interests. The state both facilitates access to education for students and limits the educational advancement of students. The state—in the form of the state government—can enhance access by providing basic education at community colleges without cost to students, as is the case in the state of North Carolina. But, the state—in the form of the federal government—can limit educational opportunities through its welfare-to-work policies, which support students for inadequate program length and then require welfare recipients to work before they have achieved required skill competencies. Adding to Carnoy's work in our understanding of the behaviors of the state with respect to institutions, such as community colleges, is the recent scholarship that critiques the neoliberal project.

Neoliberalism

One of the more influential forces for institutions over the past two decades is a political and economic ideology referred to as neoliberalism. Noam Chomsky views the term *neoliberal* as applicable to those who favor the control of social life of the many for the maximization of profit for the few.[36] Feminist scholar Catherine Kingfisher, in a critique of neoliberalism and its effects upon poverty and women's poverty in particular, notes that in the state's movement to free-market behaviors, there are attacks upon the welfare state to the extent that the responsibility for poverty devolves to the level of individuals.[37] In the neoliberal argument, welfare programs encourage dependency, and thus cuts in benefits grant freedom for individuals to pursue individual ends and realize their potential. Policies of the state, notes Kingfisher, are directed at reforming individuals, not social structures. Political economist George DeMartino argues that "neoliberalism induces inequality—domestically and globally."[38] By inequality he means economic inequality. Michael Apple summarizes the ideological commitments and ideal behaviors of neoliberalism.[39] These behaviors include, among others, the expansion of open, economic markets; the reduction of government responsibility for social needs; the reinforcement of a competitive structure for economic behaviors; and the lowering of social expectations for economic security.[40] Neoliberalism is an ideological commitment to competition, in the form of social Darwinism, to state reduction of social programs and to state support for players, especially corporations, in international markets. Neoliberalism is a political project aimed at institutional change, and education is one of those institutions where the norms are undergoing severe pressure to change.[41]

Lisa Duggan's critique of neoliberalism connects the economic and political components and actions of neoliberalism with identity politics, suggesting that it differs from earlier versions of liberalism. "Neoliberalism, a late twentieth-century incarnation of Liberalism," she writes, "organizes material and political life in terms of race, gender, and sexuality as well as economic class and nationality, or ethnicity and religion."[42] One of her targets in this critique is the poor, particularly poor women of color, and her example of the dismantling of social welfare in the United States shows that such actions lead to increasing economic inequality as money moves from the public sector to the private.

More vitriolic than Duggan is Henry Giroux, whose argument ranges from explanation to condemnation of neoliberal ideology and practice in the United States.[43] In Giroux's critique, neoliberalism is a twenty-first century parallel to William Blake's "dark Satanic mills" of dehumanizing industrialism in late-eighteenth and early-nineteenth century England. Giroux notes, "[N]eoliberalism is an ideology and politics buoyed by the spirit of a market fundamentalism that subordinates the art of democratic politics to the rapacious laws of a market economy that expands its reach to include all aspects of social life within the dictates and values of a market-driven society."[44]

Neoliberal practices lead to a weakened social state, replacing the social contract and the public good with personal responsibility and a competitive and vicious individualism. Dismantled are the New Deal policies that expand social provisions including health care, public transportation, housing, employment, unemployment benefits, and education.[45] In France, Alain Touraine equates neoliberalism with capitalism; yet, unlike Giroux, he sees that neoliberalism is destructible through a social critique.[46]

Critics of neoliberalism frame their dissent upon the subordination of the public good to individual economic interests, as well as the dominance of corporations and corporate interests over individual agency. In this critique, neoliberalism is antithetical not only to popular conceptions of social justice as well as to Rawls' theoretical view of justice, but also to fundamental principles of the community college.

Higher Education and Neoliberalism

In the latter part of the 1990s and into the beginnings of this century, there have been a spate of publications on the economic competitive orientation of higher education institutions, both nationally and internationally, such as Burton Clark's focus upon entrepreneurial institutions, Slaughter and Leslie's academic capitalism, Marginson and Considine's enterprise universities, Bok's commercialization of higher education, Levin's globalized community colleges, and most recently Slaughter and Rhoades's "academic capitalism knowledge/learning regime."[47] These are responses to growing concerns about the shift of higher education institutions internationally to behaviors that suggest a prevailing orientation to economic matters as opposed to more social or cultural endeavors. For Marginson and Considine, students have become economic entities not citizens. For Slaughter and Leslie, faculty in research universities have become independent entrepreneurs, seeking funds to support research but directed by those resources to provide research for the private sector's economic returns. And, for Bok, higher education institutions have turned to commercialization, selling the work of employees and students, whether that work is created in university laboratories or on the gridiron. In spite of the differences of higher education institutions—differences of institutional type and national location—behaviors have become isomorphic and the outcomes similar. For me, as well as for other scholars, these institutional behaviors are captured in the political concept of neoliberalism: what education scholar Nelly Stromquist calls "an economic doctrine that sees the market as the most effective way of determining production and satisfying people's needs."[48]

Education is not immune to neoliberalism's forms of governance, which include "shift[s] in the direction of increasing marketization, a redrawing of the public/private distinction, valorization of possessive individualism, and shifts in state expenditure (often accompanied by increasing state interference) in social arenas."[49] Adriana Puiggrós argues that in the public school sector, "Neoliberal educational policies subordinate democracy to the market and evaluation to control."[50] The university is more aptly named "the neoliberal university," according to Slaughter and Rhoades, who see that the social roles of public higher education have been displaced by the economic role of serving corporations. This is accomplished in two ways: first, in generating revenues and producing for a market; second, in managing institutions to reduce the power of labor.[51] The community college is more definitively named "the globalized community college." It is responsive to global flows of capital, immigration, and information as well as an instrument of the state, particularly a neoliberal state.[52]

In adjusting to an increasingly competitive market and in responding to both their traditional and new student populations, colleges and universities have adopted a decidedly business and corporate orientation to their operations.

Academics at universities have shifted behaviors to emulate capitalists, forming a web-like structure of entrepreneurs routinely pursuing money. In Australia, universities comprise a vast network of the state's apparatus to transform the economy: Universities serve as a model of entrepreneurial behaviors, with faculty and institutions competing against each other.[53] In Europe, a select group of universities function as businesses, producing goods and providing services that only vaguely match the popular conception of academic education.[54] Even more extreme, in China, universities operate shopping malls and retail stores that have no relationship to the academy or to student learning.[55]

Higher education institutions are pushed by governments and corporations, and even by higher education leaders, to conform to economic globalization, which is the vehicle of neoliberalism.[56] Higher learning has become a global business. Furthermore, the outcomes of higher education are increasingly directed to the new economy,[57] with its emphasis upon electronic technologies as critical tools of the labor force. An information-based economy requires greater numbers of more highly skilled educated employees. It also requires a repeated upgrading of skills to keep pace with the changes in the information, communications, and computer technologies being used in the technology-intensive workplace.[58] These emphases are potentially in conflict with the needs of disadvantaged students in higher education—those who require basic skills, social education, and personal attention. On one front, while educational leaders promote postsecondary education and student achievement, institutional policies and actions—from accountability measures to removal of remedial education programs from the curriculum—have negative effects on disadvantaged students. On the other front, while political leaders and policymakers champion further education for global competitiveness, government legislation, policies, or actions punish those—such as undocumented immigrants and the poor—who do not conform to either legalistic or moral strictures of the state. Indeed, the state has less difficulty spending public resources on prisons than it does on community colleges. As one community college president observed in conversation with me, "Why are we spending so much money on incarcerating people and so little money on educating them? Let's quit spending all of this money on juvenile justice and incarcerating and making new rules to incarcerate people."

Globalization

In the contemporary context, neoliberalism is associated with globalization, giving globalization a decidedly economic slant. Globalization as a process is replaced with globalization as a concept that combines international and transnational interactions and a market ideology.[59] This latter understanding of globalization identified with a single imperative—in this case, capitalism or hypercapitalism[60]—restricts the analytical potential of globalization. It omits, for example, the multiple meanings and dimensions of globalization[61] and assumes that homogenization and unity (e.g., one world system) are its outcomes.[62] The cultural domain of globalization in some distinction from the economic can suggest that plurality is more pronounced than unity, that hetero-geneity more evident than homogenization, and that local meanings trump uni-versal ones.[63] Certainly, characteristics of current trends of globalization—such as the way the movement of capital, information, and ideas is not affected by geo-graphical distances, as well as the spread of Western institutions and Western val-ues internationally and pervasive ideological conflicts that are both local and global—suggest more than economic or capitalistic behaviors and motivations.[64]

For my analysis, however, I have resorted to the more narrowly articulated view of globalization that connects the concept to capitalism and consumerism and a neoliberal agenda of both freeing markets from state controls and expand-ing markets so that those who are productive can benefit,[65] albeit with justifica-tion that the economic productivity of the few or of corporations benefits all. Similar to neoliberalism, this view absolves the state from responsibility to its cit-izens in such areas as education, welfare, health, and domestic life, among oth-ers. In my analysis of the experiences of nontraditional students, globalization has had deleterious effects: For example the North America Free Trade Agreement and the outsourcing of production to other countries have led to a loss of jobs especially for those in the manufacturing industries. These displaced workers attend community colleges in search of skills so that they can find gain-ful employment; yet, the labor market is such that even with newly acquired skills, they will land in minimum-wage or low-wage jobs in the service economy. Changes in production in the global economy as a result of technologically enhanced production and management suggest that without high-level skills and advanced education, students in community colleges who either do not move on to the baccalaureate degree or do not attain an associate's degree in a market-aligned field, such as nursing or other health science professions or business or technology-related careers, will be relegated to the underclass in society.[66]

Globalization and neoliberalism represent the utilitarian conception of jus-tice that Rawls opposes. To Rawls, national (or local) economic competitiveness cannot justify the commodification of students, in which their rights to equal-ity of opportunity are sacrificed for a larger good. This is the case even if greater competitiveness would have increased the net utility in society. That larger good becomes irrelevant once we can conclude that (a) the student does not enjoy substantive equality of opportunity in a globalized educational system and (b) the disadvantaged student is not better off as a result of the neoliberal distribu-tive scheme.

Organizational Power

Such concerns take us to the issue of organizational power. Higher education literature prior to the 1990s suggests that those who influenced institutional actions were internal decision makers, governing boards, presidents, adminis-trative executives, and in some situations faculty unions and their leaders. Brian Pusser, Sheila Slaughter, and Gary Rhoades and others have recently indicated that in the 1990s and 2000s, universities were shaped by external actors—politicians, business and industry leaders, and multinational corporations.[67] While there is limited research to identify powerful actors at community colleges, government prodded by business interests is a major player. Where government's role is limited, business and industry are major power actors.[68] In California, faculty unions are identified as one of the power actors, but this influence seems to be limited to the constraints on the management of the institution and not to the goals of the organization[69]—which Mintzberg views as central to organizational power.[70] Powerful unions in California can and do bargain for significant salary increases that move resources to employees and perhaps away from students, but there is little or no evidence to suggest that this shift in resource allocation has altered the mission and goals of the institution. In contrast, the alteration of legislation in California in the 1990s to include economic development as a major purpose of California's community colleges is a shift in mission, and thus power, in this case, was exerted by the state government and those influencers who stand to gain from education and training for economic purposes. According to community college practitioners, the economic development mission in California was the creation of the state.[71] While state government, through legislation, is a major power actor in community colleges,[72] the forces and actors who influence state legislators and the governor have significant power in the institution. The interests of these actors are not always aligned with college members, and college faculty and administrators can become unwitting agents of these actors.[73] Notwithstanding these pressures from powerful forces and actors, institutional members—administrators, faculty, and staff—do have influence on student experiences and outcomes, through actions that are not always legitimate with respect to or sanctioned by institutional or government policy, or indeed normative within the institution or consistent with the interests of external influencers. I refer to these internal actors as autonomous agents, based upon Lipsky's "street-level bureaucrats."[74] They are organizational members who counterbalance the influence of those whose interests might not be in accord with those of students, especially nontraditional students who are disadvantaged. Within the institution, these actors are administrators, faculty, and staff, as well as college presidents or chancellors. Outside the institution, these actors are executives within the system, including district and state executive officers.

Organizational power, however, is contextualized and largely beyond individuals. Indeed, the rising managerial culture of higher education institutions,[75] in part a response to unprecedented external pressures upon these institutions, has turned a collectivity into a single, seemingly monolithic enterprise. Slaughter and Rhoades refer to this situation as the "academic capitalism/knowledge learning regime," and others have referred to it as the corporatization of the academy, whereby a college or university is viewed as a single entity directed by a dominant locus of power. These pressures include government controls and requirements for accountability, resource dependence upon nongovernment organizations and businesses, new technologies, and the new economy—often captured in the concepts of the knowledge economy and globalization—and new and more strident demands from both the public and private sectors.

New Managerialism and Corporatism

The ideology of neoliberalism has infiltrated the management and governance of colleges and universities. The shift is in accord with sociologist Martin Trow's view of the governance change in universities over the past thirty years: a visible shift from "soft managerialism" to "hard managerialism." Soft managerialism is equated with collegiality and professional consensus; hard management is equated with contractual relations and autocratic control.[76] In the neoliberal context, higher education institutions have also been pushed by the state, through resource allocation (and its diminishing quantities) and accountability measures (and their increasing intrusions), to both homogenize and privatize higher education for the maximization of outputs—all in an acclaimed effort to maintain or increase access for students. The shift here is away from the goals of knowledge acquisition and free, critical, and systematic inquiry, as well as undermining the traditions of academic elitism. These alterations are consistent with the neoliberal project, promulgated by the state and corporations.[77] Terms such as "reform," "improvement," and "accountability" have become the flagwords of the movement, and increasingly information technology and systems-thinking are the symbols of its progress. The entry of quality movements or improvement initiatives into higher education is the prototypical business solution to perceived problems in the academy. These initiatives, such as total quality management (TQM), are a form of management system that relies upon continuous improvement to products and processes to create quality outputs—better products; better profits.[78] While some claim that quality management is an innovation, others, particularly academic scholars, see it as a fad or, even more disparagingly, a pernicious invasion of an antiacademic ideology. Its characteristics include standardization of production, team organization and decision making, and a clear mission conveyed by a leader. Robert Birnbaum calls it a failure in higher education because it did not take root and last.[79] Higher education scholar Estela Bensimon refers to TQM as gendered, racist, and largely exclusive. Because of its drive for standardization, TQM denies pluralism, casting gays, lesbians, African Americans, Latinos, and other nonmajority populations in the academy as outcasts, identities that need to be shaped so that they form a unit, a homogenized whole.[80] Accreditation review processes and strategic planning are two other similar organizational behaviors that suggest organizational improvement, but they too lead to a unitary system that eschews heterogeneity. The art of homogenization and standardization, of course, reduces identities, blurring differences, and subordinates individual agency. Improvement initiatives in higher education have at least two major motivations: They are a mimetic tool that demonstrates that higher education is a business, and thus has legitimacy for corporate and government stakeholders; and they are a strategy undertaken by colleges and universities in the face of demands from accountability from state governments. The positive aspect of these initiatives, theoretically, is that they carry symbolic value, showing colleges and universities as innovative, progressive, and responsive: Their use demonstrates the concern of higher education institutions for efficiency and improvement. Because colleges and universities are highly symbolic systems and because actions are loosely coupled to outcomes, change initiatives are not easily, if at all, adopted at the core of the institution.[81]

In a negative vein, we might see improvement efforts as firmly embedded in the managerial ethos of colleges and universities. Gary Rhoades and Sheila Slaughter argue that quality movements are part of an ideology that values business and industry, not students, because these entities are the actual customers—those who purchase the products, be they products of the lab or the classroom.[82] Quality improvement efforts, including review and assessment of programs, then, are tools to maximize the revenue-generating capacity of higher education institutions and reduce faculty power.[83] Ironically, quality improvement schemes and assessments—including teacher evaluations—reduce human performance to fit quantitative models and thus have a skewed vision of quality. Academic work is increasingly structured to fit the demands of, on the one hand, a machine bureaucracy and, on the other hand, a neoliberal state and its elite stakeholders.[84]

These organizational actions are aspects of what Rosemary Deem has labeled "new managerialism," which she defines as "the adoption by public sector organizations of organizational forms, technologies, management practices, and values more commonly found in the private business sector."[85] Such a condition, according to Deem, leads to the alteration of the values of the public sector employees so that they imitate or approximate those of employees in the private sector. This approach to the management of colleges and universities has been explained by scholars such as Sheila Slaughter and Gary Rhoades and vilified by Eric Gould. None of these scholars see this approach as transient: Slaughter and Rhoades refer to it as a "regime" and Gould as "corporate."[86] In this pattern of management, justice for students may not conform to the values and practices of the corporation or regime that favors what Michael Apple calls "thin morality"—where individual competitiveness is the *modus vivendi*—in distinction to "thick morality"—where principles of the common good are the basis for action.[87]

Particularly appropriate for this discussion of nontraditional students and those who are at the margins of social, economic, and educational prosperity is the recent work of Kathleen Shaw, Sara Goldrick-Rab, Christopher Mazzeo, and Jerry Jacobs on the working poor and welfare women.[88] Their research on welfare reform elucidates the influence of ideology on both policy and practice, affecting students and potential students in their access to and acquisition of education and training. They demonstrate that neoliberal ideology is indeed altering traditional understandings of educational opportunities.

In concrete ways, welfare reform and the Workforce Investment Act represent a sea change in this country's beliefs about the role that education and training should play in providing opportunities for social mobility for our most disadvantaged populations. Driven by the idea of "work first," these policies directly contradict a central tenet of American society: Instead of giving the poor opportunity to become self-sufficient by obtaining the training and education needed to lift them out of poverty, poor adults now need to enter the world of work as quickly as possible, regardless of pay, benefits, or the stability of the job. In short, higher education is not for all.[89] Through such an ideology and its impact upon policy and practice, Rawls' justice is unlikely to be realized.

The Neoliberal/Justice Conflict

If those in positions of institutional authority in higher education abide by the state's neoliberal ideology, then can students—especially disadvantaged students—receive Rawls' equality of opportunity, and can their treatment be fair? If the state controls higher education institutions and the state is a neoliberal state, then Mintzberg's power configuration of "instrument" is in play. That is, higher education institutions are extensions and vehicles of the state for promulgating and practicing neoliberal ideology.[90] College administrators and faculty are simply working for the state and are complicit in unfair practices that may include the underfunding of specific programs in relation to other programs, denying equal access to the institution because of an individual's residency status, and the withdrawal of services such as tutoring because of a revocation of a federal or state program. These actions are particularly devastating in the community college, where there is a substantial population that historically, as groups or individuals, has experienced an unequal distribution of values such as liberty and opportunity, income and wealth, and self-respect.

Theoretically, we can expect that the community college is the educational site where the conflict between neoliberalism and justice is played out and either resolved or not. Certainly since the 1970s, the community college has assumed the role of the open-access, multipurpose, and socially democratizing institution.[91] However, the institution, particularly since the 1980s, has adopted a more business-like approach, pursuing revenues, working for increased productivity, and marketing itself as a salvation for local and even state and national economies through economic development.[92] Arguably, the institution has not only framed itself as a neoliberal college, but also acted to support that label.[93]

In practice, this situation has led to an internally conflicted environment, including the apparent paradox of executives who manage the neoliberal policies but who, on a personal level, reject them. The conflict, in short and in general, is between social democratic principles and a consumer-based approach to education.[94] The goals and actions of community colleges include responding to the demands and fulfilling the needs of all members of its communities, including disadvantaged populations; and policies—whether state funding behaviors or business and industry arrangements with colleges—favor attention to economic and private interests and outcomes. Thus, college system chancellors and presidents, in their roles and policy mandates, give precedence to such matters as state economic development—providing contract training to businesses—but articulate the needed attention to disconnected youth. Within the institution, there are numerous behaviors that suggest that practitioners act outside of their roles or mandates, either ignoring neoliberal policies or placing them in a subordinate role to human and social services.

Notes

1. Arthur Cohen and Florence Brawer, *The American Community College* (San Francisco: Jossey-Bass, 2003); Kevin Dougherty, *The Contradictory College* (Albany: State University of New York Press, 1994); John Frye, "Educational Paradigms in the Professional Literature of the Community College," in *Higher Education: Handbook of Theory and Research*, ed. John Smart, 181–224 (New York: Agathon, 1994); W. Norton Grubb and Marvin Lazerson, *The Education Gospel* (Cambridge, MA: Harvard University Press, 2004); David Labaree, "Public Goods, Private Goods: The American Struggle over Educational Goals," *American Educational Research Journal* 34, no. 1 (1997): 39–81.

2. Dougherty, *The Contradictory College*; W. Norton Grubb, *Honored but Invisible: An Inside Look at Teaching in Community Colleges* (New York: Routledge, 1999); Ken Kempner, "The Community College as a Marginalized Institution" (unpublished paper presented at annual meeting of Association of the Study of Higher Education, Boston, 1991); Tania Levey, "Reexamining Community College Effects: New Techniques, New Outcomes"(paper presented at the Association for the Study of Higher Education, Philadelphia, November 2005).

3. See the publication *The Community College Journal* for examples.

4. "College Enrollment by Racial and Ethnic Group, Selected Years," in *The Chronicle of Higher Education: Almanac Issue, 2005–2006* (Washington, DC: Chronicle of Higher Education, 2005).

5. Kent A. Phillippe and Leila Gonzalez Sullivan, *National Profile of Community Colleges: Trends and Statistics*, 4th ed. (Washington, DC: American Association of Community Colleges, 2005).

6. Ernest T. Pascarella and Patrick Terenzini, *How College Affects Students: A Decade of Research* (San Francisco: Jossey-Bass, 2005).

7. Simon Marginson and Mark Considine, *The Enterprise University: Power, Governance, and Reinvention in Australia* (New York: Cambridge University Press, 2000); Sheila Slaughter and Gary Rhoades, *Academic Capitalism and the New Economy: Markets, State, and Higher Education* (Baltimore: Johns Hopkins University Press, 2004).

8. Deborah L. Floyd, Michael L. Skolnik, and Kenneth P. Walker, eds., *The Community College Baccalaureate: Emerging Trends and Policy Issues* (Sterling, VA: Stylus, 2004); John Levin, "The Community College as a Baccalaureate-Granting Institution," *The Review of Higher Education* 28, no. 1 (2004): 1–22; John S. Levin, "The Higher Credential" (Tucson, AZ: Canadian Embassy in Washington, DC, 2001).

9. Grubb and Lazerson, *The Education Gospel*; Alisa Nagler, "The Impact of Community College Education on the Nursing Shortage Crisis" (Raleigh: Department of Adult and Higher Education, North Carolina State University, 2005).

10. Slaughter and Rhoades, *Academic Capitalism*.

11. Thomas R. Bailey and Vanessa Smith Morest, "The Organizational Efficiency of Multiple Missions for Community Colleges" (New York: Teachers College,

Columbia University, 2004); John Levin, "The Business Culture of the Community College: Students as Consumers; Students as Commodities," in *Arenas of Entrepreneurship: Where Nonprofit and For-Profit Institutions Compete*, ed. Brian Pusser, special issue, *New Directions for Higher Education* 129:11–26; John Levin, *Globalizing the Community College: Strategies for Change in the Twenty-First Century* (New York: Palgrave, 2001).

12. John Levin, Susan Kater, and Richard Wagoner, *Community College Faculty: At Work in the New Economy* (New York: Palgrave, 2006); Slaughter and Rhoades, *Academic Capitalism*; Richard. L Wagoner, Amy Scott Metcalfe, and Israel Olaore, "Fiscal Reality and Academic Quality: Part-Time Faculty and the Challenge to Organizational Culture at Community Colleges," *Community College Journal of Research and Practice* 29 (2005): 1–20.

13. John S. Levin, "Neo-liberal policies and community college faculty work," in *Handbook of Higher Education*, vol. 22, ed. John Smart and William Tierney 451–496 (Norwell, MA: Kluwer Academic Publishers, 2007).

14. Richard Alfred and Patricia Carter, "New Colleges for a New Century: Organizational Change and Development in Community Colleges," in *Higher Education: Handbook of Theory and Research*, ed. John C. Smart and William G. Tierney, 240–83 (New York: Agathon, 1999); Bailey and Morest, "Organizational Efficiency"; Anthony P. Carnevale and Donna M. Desrochers, *Help Wanted . . . Credentials Required: Community Colleges in the Knowledge Economy* (Washington, DC: Educational Testing Service and the American Association of Community Colleges, 2001); Anthony P. Carnevale and Donna M. Desrochers, "Why Learning? The Value of Higher Education to Society and the Individual," in *Keeping America's Promise* (Denver: Education Commission of the States, 2004), 39–43; Kevin J. Dougherty and Marianne Bakia, "The New Economic Role of the Community College: Origins and Prospects" (occasional paper, Community College Research Center, Teachers College, New York, June 1998); John E. Roueche and Suanne D. Roueche, *High Stakes, High Performance: Making Remedial Education Work* (Washington, DC: Community College Press, 1999); Roueche, Lynn Taber, and Roueche, *The Company We Keep: Collaboration in the Community College* (Washington, DC: American Association of Community Colleges, 1995).

15. D. Franklin Ayers, "Neoliberal Ideology in Community College Mission Statements: A Critical Discourse Analysis," *The Review of Higher Education* 28, no. 4 (2005): 527–49;); Alicia C. Dowd, "From Access to Outcome Equity: Revitalizing the Democratic Mission of the Community College," in "Community Colleges: New Environments, New Directions," ed. Kathleen Shaw and Jerry Jacobs, special issue, *The ANNALS of the American Academy of Political and Social Science* 586 (March 2003): 92–119; Shaw et al., "Putting Poor People to Work: How the Work-First Ideology Eroded College Access for the Poor" (unpublished manuscript, 2005); Shaw and Sara Rab, "Market Rhetoric versus Reality in Policy and Practice: The Workforce Investment Act and Access to Community College Education and Training," *The ANNALS of the American Academy of Political and Social Science* (March 2003).

16. Michael Apple, "Comparing Neoliberal Projects and Inequality in Education," *Comparative Education* 37, no. 4 (2001): 409–23; Richard Bagnall, "Lifelong Learning and the Limitations of Economic Determinism," *International Journal of Lifelong Education* 19, no. 1 (2000): 20–35; Bailey and Morest, "Organizational Efficiency"; Catherine Casey, *Work, Society and Self: After Industrialism* (New York: Routledge, 1995); Rosemary Deem, "'New Managerialism' and Higher Education:

The Management of Performances and Cultures in Universities in the United Kingdom," *International Studies in Sociology of Education* 8, no. 1 (1998): 47–70; Jim Gallacher et al., "Learning Careers and the Social Space: Exploring the Fragile Identities of Adult Returners in the New Further Education," *International Journal of Lifelong Education* 21, no. 6 (2002): 493–509; Mark Olssen, "The Neoliberal Appropriation of Tertiary Education Policy: Accountability, Research, and Academic Freedom" (2000), cited May 2004, available at http://www.surrey.ac.uk/Education/profiles/olssen/neo-2000.htm; Andrew Ross, *No-Collar: The Humane Workplace and Its Hidden Costs* (New York: Basic Books, 2003); Carlos A. Torres and Daniel Schugurensky, "The Political Economy of Higher Education in the Era of Neoliberal Globalization: Latin America in Comparative Perspective," *Higher Education* 43 (2002): 429–55; Roger Waldinger and Michael I. Lichter, *How the Other Half Works: Immigration and the Social Organization of Labor* (Berkeley: University of California Press, 2003); Shirley Walters and Kathy Watters, "Lifelong Learning, Higher Education, and Active Citizenship: From Rhetoric to Action," *International Journal of Lifelong Education* 20, no. 6 (2001): 471–78.

17. John Rawls, *A Theory of Justice* (Cambridge, MA: Belknap Press of Harvard University Press, 1999).

18. Rawls, *Political Liberalism* (New York: Columbia University Press, 1993).

19. Rawls, *A Theory of Justice.*

20. William Bowen and Derek Bok, *The Shape of the River: Long-Term Consequences of Considering Race in College and University Admissions* (Princeton, NJ: Princeton University Press, 1998).

21. Quentin Bogart, "The Community College Mission," in *A Handbook on the Community College in America*, ed. George Baker, 60–73 (Westport, CT: Greenwood, 1994); Steven Brint and Jerome Karabel, *The Diverted Dream: Community Colleges and the Promise of Educational Opportunity in America, 1900–1985* (New York: Oxford University Press, 1989); Cohen and Brawer, *The American Community College*; Grubb, *Honored but Invisible*; Richard Richardson and Louis Bender, *Fostering Minority Access and Achievement in Higher Education* (San Francisco: Jossey-Bass, 1987); Roueche and George A. Baker III, *Access and Excellence* (Washington, DC: Community College Press, 1987); John E. Roueche and Suanne D. Roueche, *Between a Rock and a Hard Place: The At-Risk Student in the Open-Door College* (Washington, DC: American Association of Community Colleges, 1993); Kathleen Shaw, Robert Rhoads and James Valadez, eds., *Community Colleges as Cultural Texts* (Albany: State University of New York Press, 1999).

22. David Held et al., *Global Transformations: Politics, Economics, and Culture* (Stanford, CA: Stanford University Press, 1999); Gary Teeple, *Globalization and the Rise of Social Reform* (Atlantic Highlands, NJ: Humanities Press, 1995); Malcolm Waters, *Globalization* (New York: Routledge, 1996).

23. George DeMartino, *Global Economy, Global Justice: Theoretical Objections and Policy Alternatives to Neoliberalism* (New York: Routledge, 2000).

24. Alexander Astin and Leticia Oseguera, "The Declining 'Equity' of American Higher Education," *The Review of Higher Education* 27, no. 3 (2004): 321–41; Bowen and Bok, *The Shape of the River*; Douglas S. Massey et al., *The Source of the River: The Social Origins of Freshmen at America's Selective Colleges and Universities* (Princeton, NJ: Princeton University Press, 2003); K. Edward Renner, "Racial Equity and Higher Education," *Academe* (January–February 2003): 38–43.

25. Levin, *Globalizing the Community College.*

26. Sheila Slaughter and Larry Leslie, *Academic Capitalism, Politics, Policies, and the Entrepreneurial University* (Baltimore: Johns Hopkins University Press, 1997).

27. Derek Bok, *Universities in the Marketplace: The Commercialization of Higher Education* (Princeton, NJ: Princeton University Press, 2003); Bill Readings, *The University in Ruins* (Cambridge, MA: Harvard University Press, 1997); Murray Sperber, *Beer and Circus: How Big-Time Sports Is Crippling Undergraduate Education* (New York: Henry Holt, 2000).
28. Dougherty, *The Contradictory College.*
29. Paul DiMaggio and William Powell, Introduction to *The New Institutionalism in Organizational Analysis*, ed. William Powell and Paul DiMaggio, 1–40 (Chicago: University of Chicago Press, 1991); W. Richard Scott, *Institutions and Organizations* (Thousand Oaks, CA: Sage, 1995).
30. Yeheskel Hasenfeld, *Human Service Organizations* (Englewood Cliffs, NJ: Prentice Hall, 1983).
31. Penelope. E. Herideen, *Policy, Pedagogy, and Social Inequality: Community College Student Realities in Postindustrial America* (Westport, CT: Bergin and Garvey, 1998).
32. Rawls, *Political Liberalism*; Rawls, *A Theory of Justice*; John Ralston Saul, *The Unconscious Civilization* (Concord, ON: House of Anansi Press, 1995); Charles Taylor, *The Ethics of Authenticity* (Cambridge, MA: Harvard University Press, 1991); Teeple, *Globalization and the Rise of Social Reform.*
33. Henry Mintzberg, *Power in and Around Organizations* (Englewood Cliffs, NJ: Prentice Hall, 1983).
34. Martin Lipsky, *Street-Level Bureaucracy* (New York: Russell Sage Foundation, 1980).
35. Martin Carnoy, *The State and Political Thought* (Princeton, NJ: Princeton University Press, 1984).
36. Noam Chomsky, *Profit Over People: Neoliberalism and Global Order* (New York: Seven Stories Press, 1999).
37. Catherine Kingfisher, *Western Welfare in Decline: Globalization and Women's Poverty* (Philadelphia: University of Pennsylvania Press, 2002).
38. DeMartino, *Global Economy, Global Justice.*
39. Apple, "Comparing Neoliberal Projects."
40. Ibid.
41. John Campbell and Ove Pedersen, "Introduction: The Rise of Neoliberalism and Institutional Analysis," in *The Rise of Neoliberalism and Institutional Analysis*, ed. John Campbell and Ove Pedersen, 2–23 (Princeton, NJ: Princeton University Press, 2001).
42. Lisa Duggan, *The Twilight of Equality? Neoliberalism, Cultural Politics, and the Attack on Democracy* (Boston: Beacon Press, 2003).
43. Henry Giroux, *The Terror of Neoliberalism* (Boulder, CO: Paradigm, 2004).
44. Ibid., xxii.
45. Ibid.
46. Alain Touraine, *Beyond Neoliberalism*, trans. David Macey (Malden, MA: Blackwell, 2001).
47. Bok, *Universities in the Marketplace*; Burton Clark, *Creating Entrepreneurial Universities: Organisational Pathways of Transformation* (Oxford: Pergamon, 1998); Levin, *Globalizing the Community College*; Marginson and Considine, *The Enterprise University*; Slaughter and Leslie, *Academic Capitalism, Politics, Policies*; Slaughter and Rhoades, *Academic Capitalism.*
48. Nelly P. Stromquist, *Education in a Globalized World: The Connectivity of Economic Power, Technology, and Knowledge* (Lanham, MD: Rowman and Littlefield, 2002).

49. Kingfisher, *Western Welfare in Decline.*
50. Adriana Puiggrós, *Neoliberalism and Education in the Americas* (Boulder, CO: Westview, 1999).
51. Slaughter and Rhoades, "The Neoliberal University," *New Labor Forum* (Spring/Summer 2000): 73–79.
52. Levin, *Globalizing the Community College.*
53. Simon Marginson, *Educating Australia: Government, Economy, and Citizen Since 1960* (Melbourne: Cambridge University Press, 1997); Marginson and Considine, *The Enterprise University.*
54. Clark, *Creating Entrepreneurial Universities.*
55. Observation made by Anning Ding, doctoral student at the University of Arizona and Chinese citizen, in personal conversation with this author, 2000.
56. Jan Currie and Janice Newson, eds., *Universities and Globalization* (Thousand Oaks, CA: Sage, 1998); Slaughter, "Who Gets What and Why in Higher Education? Federal Policy and Supply-side Institutional Resource Allocation" (presidential address, Association for the Study of Higher Education annual meeting, Memphis, TN, 1997); Slaughter and Leslie, *Academic Capitalism, Politics, Policies.*
57. Grubb and Lazerson, *The Education Gospel.*
58. Slaughter and Rhoades, *Academic Capitalism.*
59. Jan Currie, "Introduction to *Universities and Globalization,* ed. Jan Currie and Janice Newson, 1–13 (Thousand Oaks, CA: Sage, 1998); Noel Gough, "Globalization and Curriculum: Theorizing a Transnational Imaginary" (paper presented at the Annual Meeting of the American Educational Research Association, San Diego, 1998); Held et al., *Global Transformations.*
60. James Paul Gee, Glynda Hull, and Colin Lankshear, *The New Work Order: Behind the Language of the New Capitalism* (Boulder, CO: Westview, 1996); Gordon Laxer, "Social Solidarity, Democracy, and Global Capitalism," *The Canadian Review of Sociology and Anthropology* (August 1995): 287–312.
61. Arjun Appadurai, "Disjunctures and Difference in the Global Cultural Economy," in *Global Culture: Nationalism, Globalization, and Modernity,* ed. Mike Featherstone, 295–310 (Newbury Park, CA: Sage, 1990); Roland Robertson, *Globalization: Social Theory and Global Culture* (London: Sage, 1992).
62. Janice Dudley, "Globalization and Education Policy in Australia," in *Universities and Globalization,* ed. J. Currie and J. Newson, 21–43 (Thousand Oaks, CA: Sage, 1998).
63. Johann Arnason, "Nationalism, Globalization, and Modernity," in *Global Culture: Nationalism, Globalization, and Modernity,* ed. Featherstone, 207–36 (Newbury Park, CA: Sage, 1990); Appadurai, "Disjunctures and Difference"; Scott Davies and Neil Guppy, "Globalization and Educational Reforms in Anglo-American Democracies," *Comparative Education Review* 41, no. 4 (1997): 435–59.
64. Richard De Angelis, "Globalization and Recent Higher Education Reforms in Australia and France: Different Constraints; Differing Choices in Higher Education Structure, Politics, and Policies" (paper prepared for the 9th World Congress on Comparative Education, Sydney, Australia, July 1997); Richard Edwards and Robin Usher, "Globalization, Diaspora Space, and Pedagogy" (paper presented at the annual meeting of the American Educational Research Association, San Diego, April 1998); Gough, "Globalization and Curriculum"; David Held and Anthony McGrew, "Globalization and the Liberal Democratic State," *Government and Opposition: An International Journal of Comparative Politics* 23, no. 2 (1993): 261–88; E. Fuat Keyman, *Globalization, State, Identity, and Deference* (Atlantic Highlands, NJ: Humanities Press, 1997).

65. Gary Teeple, *Globalization and the Decline of Social Reform* (Atlantic Highlands, NJ: Humanities Press, 1995).

66. Manuel Castells, *The Rise of the Network Society*, 2nd ed. (Malden, MA: Blackwell, 2000); W. Norton Grubb, "Learning and Earning in the Middle: The Economic Benefits of Sub-baccalaureate Education" (occasional paper, Community College Research Center, Teachers College, New York, September 1998); Grubb and Lazerson, *The Education Gospel*; Jeremy Rifkin, *The End of Work: The Decline of the Global Labor Force and the Dawn of the Postmarket Era* (New York: G. P. Putnam's Sons, 1995).

67. Brian Pusser, "Beyond Baldrige: Extending the Political Model of Higher Education Organization and Governance," *Educational Policy* 17, no. 1 (2003): 121–40; Brian Pusser, *Burning Down the House: Politics, Governance, and Affirmative Action at the University of California* (Albany: State University of New York Press, 2004); Slaughter and Rhoades, *Academic Capitalism*; Slaughter and Rhoades, "The Neoliberal University."

68. Dougherty and Bakia, "The New Economic Role"; Levin, "The Business Culture"; Levin, "The Community College"; Levin, *Globalizing the Community College*; Levin, "Public Policy, Community Colleges, and the Path to Globalization," *Higher Education* 42, no. 2 (2001): 237–62; Ken Meier, "The Community College Mission and Organizational Behavior" (unpublished paper, The Center for the Study of Higher Education, Tucson, AZ, 1999); Susan Twombly and Barbara Townsend, "Conclusion: The Future of Community Policy in the Twenty-First Century," in *Community Colleges: Policy in the Future Context*, ed. Barbara Townsend and Susan Twombly, 283–98 (Wesport, CT: Ablex, 2001).

69. Academic Senate for California Community Colleges, "Scenarios to Illustrate Effective Participation in District and College Governance" (Community College League of California and the Academic Senate for California Community Colleges, 1996), cited June 23, 2005, available at http://www.academicsenate.cc.ca.us/Publications/Papers/FinalScenario.htm; Karen Sue Grosz, The Hierarchical Approach to Shared Governance (Academic Senate for California Community Colleges, 1988), cited June 23, 2005, available at http://www.academicsenate.cc.ca.us/Publications/Search/originalresults.asp; Levin, *Globalizing the Community College*; Kenneth White, "Shared Governance in California," *New Directions for Community Colleges* 102 (1998): 19–29.

70. Mintzberg, *Power in and Around Organizations*.

71. Levin, *Globalizing the Community College*.

72. Sue Kater and John Levin, "Shared Governance in Community Colleges in the Global Economy," *Community College Journal of Research and Practice* 29, no. 1 (2005): 1–24.

73. John Levin, Susan Kater, and Richard Wagoner, *Community College Faculty: At Work in the New Economy* (New York: Palgrave, forthcoming).

74. Lipsky, *Street-level Bureaucracy*.

75. Deem, "'New Managerialism'"; Gee, Hull, and Lankshear, *The New Work Order*; Levin, Kater, and Wagoner, *Community College Faculty*; Gary Rhoades and Sheila Slaughter, "Academic Capitalism, Managed Professionals, and Supply-Side Higher Education," *Social Text* 15, no 2 (1997): 9–38; Slaughter and Rhoades, *Academic Capitalism*; Stromquist, *Education in a Globalized World*.

76. Martin Trow, "Managerialism and the Academic Profession: The Case of England," *Higher Education Policy* 7, no. 2 (1994): 11–18.

77. Stanley Aronowitz and William Di Fazio, *The Jobless Future: Sci-Tech and the Dogma of Work* (Minneapolis: University of Minnesota Press, 1994); Manuel Castells, *The Rise of the Network Society* (Cambridge, MA: Blackwell, 1996).

78. Lawrence Jauch and Robert Orwig, "A Violation of Assumptions: Why TQM Won't Work in the Ivory Tower," *Journal of Quality Management* 2, no. 2 (1997): 279–93.

79. Robert Birnbaum, *Management Fads in Higher Education: Where They Come from, What They Do, Why They Fail* (San Francisco: Jossey-Bass, 2000).

80. Estela Bensimon, "Total Quality Management in the Academy: A Rebellious Reading," *Harvard Educational Review* 65, no. 4 (1995): 593–611.

81. Birnbaum, *Management Fads*; Birnbaum, "The Latent Organization Functions of the Academic Senate: Why Senates Do Not Work but Will Not Go Away?" *The Journal of Higher Education* 60, no. 4 (1989): 224–43; Karl Weick, "Educational Organizations as Loosely Coupled Systems," *Administrative Science Quarterly* 21 (1976): 1–19.

82. Rhoades and Slaughter, *Academic Capitalism*.

83. Slaughter and Rhoades, "The Neoliberal University."

84. John Levin, Sue Kater, Cristie Roe, and Rick Wagoner, "Not Professionals? Community College Faculty in the New Economy" (presented at the symposium for the annual meeting of the American Educational Research Association, Chicago, April 2003); John Saul, *The Unconscious Civilization* (Concord, ON: House of Anansi Press, 1995); Slaughter, "Who Gets What and Why."

85. Deem, "'New Managerialism,'" 47.

86. Eric Gould, *The University in a Corporate Culture* (New Haven, CT: Yale University Press, 2003); Slaughter and Rhoades, *Academic Capitalism*.

87. Apple, "Comparing Neoliberal Projects."

88. Shaw et al., "Putting Poor People to Work."

89. Ibid., 191.

90. Carnoy, *The State and Political Thought*.

91. Bogart, "The Community College Mission"; Cohen and Brawer, *The American Community College*; John Dennison and Paul Gallagher, *Canada's Community Colleges* (Vancouver: University of British Columbia Press, 1986); Levin, *Globalizing the Community College*; Richard Richardson, Elizabeth Fisk, and Morris Okun, *Literacy in the Open-Access College* (San Francisco: Jossey-Bass, 1983); Roueche, Eileen E. Ely, and Roueche, *In Pursuit of Excellence: The Community College of Denver* (Washington, DC: Community College Press, 2001); George Vaughan, *The Community College Story* (Washington, DC: American Association of Community Colleges, 2000).

92. W. Grubb et al., *Workforce, Economic, and Community Development: The Changing Landscape of the Entrepreneurial Community College* (Berkeley: National Center for Research in Vocational Education, University of California, 1997); Levin, *Globalizing the Community College*; Shaw and Rab, "Market Rhetoric"; Lynn Taber, "Chapter and Verse: How We Came to Be Where We Are," in *The Company We Keep: Collaboration in the Community College*, ed. Roueche, Taber, and Roueche, 25–37 (Washington, DC: American Association of Community Colleges, 1995).

93. John Levin, "Nouveau College: Community Colleges in the New Economy" (paper presented at the American Educational Research Association, San Francisco, April 2006); Levin, Kater, and Wagoner, *Community College Faculty*.

94. Labaree, "Public Goods, Private Goods."

Chapter 3

Multiple Identities, Multiple Motivations and Goals, and the Student-Institution Disconnection

This chapter is an empirically based discussion of the identities of nontraditional students, bringing together data and findings from the examination of thirteen specific sites to illuminate where students experience both connection and lack of connection to the community college. It has three main parts: 1) the multiple identities of nontraditional community college students, 2) the motivations and goals of nontraditional students, and 3) the student-institution disconnection. The discussion uses interviews with organizational members and students at community colleges for both specific grounding and illumination.

The Identity Issue

Categories and classifications of students as a population—as male or female; as African American, Native American, Latino, or white; as over or under twenty-four; and the like—while useful for a variety of purposes, including research, neither capture the complexity of individual (and indeed group) identity nor explain college experience or performance. I begin with the assumption that "college" for students cannot be viewed or understood as the principal or primary community or "lifeworld." Indeed, college is situated within the life experiences of students and the environments they inhabit, as well as the community with which they interact on a daily basis. This condition combines both the "figured worlds" of Dorothy Holland and her colleagues, as well as the "imagined communities" of Benedict Anderson.[1] According to Holland, "Figured worlds take shape within and grant shape to the coproduction of activities, discourse, performances, and artifacts."[2] People are recruited to or enter into these worlds where there is common, agreed-upon, or negotiated understandings and interpretations of meanings. It is within the context of these worlds that individuals understand and develop at least a part of their identities. Through common participation in activities with others, individuals gain a sense of commonality or membership in a categorical social body, which Anderson terms "imagined community"—a manufactured community. This identification with an "imagined community" is particularly apparent for adult students whose association with college may be minimal or at least divided.

"Student" is but one of the components of the identity of those who attend college. Other characteristics, such as mother, child, lesbian, Muslim, and the like, also feature in this identity. Furthermore, identities are negotiated, cocreated, situated, and socially constructed, involving not only one's understanding of oneself but also a social discourse about oneself.[3] Identity is both ascribed and self-articulated—including what others impose upon individuals as well as what one says about oneself—and actual and designated—including what we know about individuals as well as what we expect from individuals.[4] Identity is embedded in discourse, in articulations about how one is viewed and views oneself.

For students, identity is a context for our understanding not only student behaviors but also student motivations and goals. Furthermore, student experiences in college both challenge and reinforce the identities students bring to college. The world of college can reshape student identities, or students can reject or ignore this world and its attributes because its challenges to identity are threatening to one's sense of self or because its practices are not consistent with a more pronounced social and personal identity. That is, the identification of students with college can be fleeting or become central to their personal life. Such is the case for all college students as amplified in recent empirical accounts[5] and fictional accounts.[6] Arguably, the identities of nontraditional students are more multifaceted or complex than traditional students.

The Multiple Identities of Nontraditional Community College Students

Nontraditional students have specific identity characteristics, such as minority status, low economic status, or immigrant status, as well as a host of other characteristics such as gender, age, parental and occupational status, and physical and mental ability/disability. Additionally, they possess identifiable traits and exhibit certain behaviors, such as limited academic achievement in previous schoolwork and lack of self-confidence.[7] Their complex backgrounds are expressed through the observations of an instructor who teaches Adult Basic Education at Johnston Community College in North Carolina.

> Sometimes I feel, when I get up in the morning and I have my class to look forward to and I'm happy and I'm upbeat and I come to work, sometimes I think about if I were in their shoes; that in some of their shoes, I probably would not even bother. Because they don't have that much to really look forward to or, you know, they don't have that much encouragement.

His personal and professional experiences with these students illuminate their psychological makeup.

> A couple of years ago I had this young guy in my class. He was so shy, and I remember in English 3 . . . I kept saying, you have to write this essay. And

I would try to help him, and one day he started shaking. He said, "All through school, I was told I could not write," and he said, "I'm afraid to try to write because I don't want to disappoint you." I said, "Oh, just forget about that." I said, "Just take this topic and think about it, and you go home tonight and you write an essay, and I'll help you with it." I said, "No matter what you do, it'll be a start. We'll take what you've got and we'll work on it." He came back the next day; he had written one of the best essays I had read in years. But all through school he'd been told that he couldn't write. . . . I'm all over the classroom and at any one time I know what just about everybody's doing. But sometimes I feel like I've been walking around eggs and eggshells all day. Because, you have to, with all of their backgrounds and all the diverse experiences they've had, you have to be very acute. If a student comes in and they're sitting there and they're not participating, and you just have a feeling, you just have to know to leave them alone that day because you don't know what's happened in their environment that's causing them to be that way; or if they forgot to take their Ritalin or just what happened. . . . [I]t can be tense at times. (Bill, Adult Basic Education Instructor, Johnston Community College)

The social and economic backgrounds of college students, as well as their previous school experiences, are not only contributors to personal identity but also affect both college experiences and college performance. In Dallas, Texas, at Mountain View College, an academic administrator characterizes his college's students. His conceptualization encompasses the entire college student body.

Their difficulties are financial . . . cultural. . . . [W]hen this college first opened it was 97% Anglo, but what happened was the demographic mix changed in thirty-three years. [Now] there is 37% Hispanic, 35% African American. . . . Our faculty base stayed the same. . . . We get a group of students who come in . . . who [previously are used to seeing] people who look like them, act like them, that are all around them . . . and here it is . . . totally different. . . . We have to continually let our faculty understand. . . . We may have to hold hands and let students understand that it is OK, that this is OK; this environment is OK. . . . Our students need a lot of support. . . . Once they get past that first hump, then we have them. But we lose a lot after that first semester. . . . Most [of our students] are nontraditional.

Here, students' backgrounds, which include how they experience the world before college, shape their perceptions and affect their experience of college. In this case at Mountain View College, the institution is discordant with student identity.

Community college students—the majority of whom are nontraditional—possess and acknowledge identities that are more than essentialist. These identities are voluntarily, assumed not ascribed, and are "multifaceted, multisourced, and multilayered."[8] Indeed, these students are members of multiple communities.[9] Dishon is a student at the Community College of Denver who is blind. Not once in his hourlong interview, accompanied by his canine companion, Franklin, does he refer to himself as a student with learning difficulties or someone who is disadvantaged because he is a returning adult student who dropped out of college. Nor does he ever say that he is African American. All of which are no doubt part of his identity, his background, and his reality. Yet, although others would perceive Dishon as both African American and disabled, assuming that his disability is both an essential part of his identity and highly disadvantageous to him and his life's prospects, Dishon identifies himself only as a student who is blind.

> Well, I just turned twenty-eight in March and I'm sort of a, well, I am a military brat. Yeah. . . . I was born in California . . . [and] I have a twin brother who lives in Louisiana now. [With] my dad, we moved around from different states, from California, to Oklahoma, to Louisiana . . . where we settled down, and I went to college there for some time, and then I decided to [go to] New Mexico on my own, just to get away from the parents. Because I wanted to be my own person, I wanted to be independent because they were doing too much for me. And before I moved to New Mexico I went to a blind center, teaching people who are newly blinded or that are blind how to cope with, how to be independent, more or less.

Indeed, Dishon works part-time fifteen hours a week as a disability advocate for the local bus system.

> I work right now for RTD, which is the regional transportation district. . . . And my job there is to, it's more ADA [Americans with Disabilities Act] compliance type stuff. It's making sure that the bus drivers, because it's by law . . . call out stops for the blind people and help make sure the people in wheelchairs get on safely and make sure they're tied down, and stuff like that. So basically what I do is—which works out for me—I ride the bus, because I ride the bus everywhere I go anyway, so I ride the bus just to make sure they're doing their job.

Furthermore, Dishon owns a car, even though, of course, he does not drive; instead, his friends drive him in his car, giving him mobility.

> It kind of works out when I have a roommate who has a license or whatever; and since I got rid of my old roommate, he's no longer there. So it's kind of whoever I can get to drive me to the grocery store. Because it's easier, I tell you; waiting on cabs here is just crazy. Sometimes I have to wait up to four hours for a cab and then I don't want to get on the bus with ten bags of groceries. So it's just easier, more convenient for me. People are like, well, why do you have your own car? You can't see to drive it. I say, well, I can't see to drive but somebody else can. (Dishon, student, Community College of Denver)

Dishon is not only resourceful, or resilient, locating creative ways in his role as a person without sight to function; but he is also altruistic, serving as an advocate for others with disabilities. His identity as a student is simply one component of a larger identity, filled with multiple characteristics.

Guillean is a nursing student at Bakersfield College in California. On top of her full-time program, she works over thirty hours a week, mostly as a janitor, cleaning offices.

> I'm Guillean. I came to Bakersfield, like, seven years ago. I came here from Mexico. I came from Mexico and I finished high school here, my last year of high school, so that was kind of hard. And I wasn't planning on coming to college right after because I wasn't, the change and everything was just a little too much for me. But I was convinced, and so one of the counselors convinced me, so I came here and I started off doing criminal justice. And my mom used to work at a nursing facility and I went over to volunteer a few times, and I kind of looked at what the nurses were doing and I thought, hmm, I like that, I think I can do that, so I changed my major and I started doing nursing.

But as a student whose second language is English and first language is Spanish, Guillean's national identity is not clear-cut: Born in the United States with formative years spent in Mexico, she is the product of two national cultures.

> I was born here. And my mom got her citizenship here; my dad as well. I struggled because of the English, because I was in Mexico for such a long time and I had to catch up with my English, my math, and all of that. (Guillean, student, Bakersfield College)

The sociocultural and economic backgrounds of all college students influence both their experiences of and performance in higher education. Access to higher education is often predicated not just on students' academic credentials, but on the amount of cultural and social capital they possess.[10] For example, among African American and Hispanic students, cultural capital and social capital have as much influence on college decision making as does student academic background.[11] Cultural capital is defined as the worldview and accumulated knowledge that middle- and upper-class families share with their offspring in order to perpetuate class and privilege throughout current and subsequent generations.[12] High school students from families without cultural capital know little about higher education opportunities available to them and the rigors and responsibilities associated with those opportunities. Social capital, in some distinction, is the means by which institutional agents, or those individuals with access to socially valuable resources and information, form networks and relationships whereby they may transmit that information to maintain an established social order.[13] A lack of social capital impairs students' access to educational opportunities.

Thus, consistent with other disadvantaged populations, specific minority student groups are disadvantaged because of their social and cultural backgrounds in that they are viewed as deficient in the necessary social and cultural capital for academic achievement. Their backgrounds predispose them to enroll in specific kinds of institutions and in specific types of programs. Latino students tend to come from lower socioeconomic backgrounds and generally choose institutions that are close to home and offer programs that help them transition to four-year institutions.[14] First-generation students—those who are first in their family to attend college—tend to perform less well than other students in college. As first-generation college students, they lack both cultural and social capital, and college is an unfamiliar environment. These students are more likely to be classified as minorities who come from low-income families.[15] Stacey at the Community College of Denver is one such student whose student experiences are shaped by her sociocultural background. Stacey's family has no educational legacy to pass on to her, and as a low-income student, she has family obligations that not only contribute to her identity but also shape how she experiences college.

> My name is Stacey, and I am an older student. I'm in my thirties now, but I started college in '96 before I had a child. So I've kind of been going to school off and on while working and supporting the child I have. I'm slowly trying to adjust to letting go of work and trying to find other means of focusing on my education, so I can get good grades and get through college so I can get my degree. . . . My mom and dad didn't get college educations. [T]hey were from the older generation, where you would find a job and just kind of stay there forever, and they don't really understand college so they don't really understand what I'm doing half the time. I'm just really trying to work on [my studies]. I live right across the street [from the college]. I just moved into a housing program to help me be able to afford rent in Denver while I'm going to college, which really helps. I've lived here all my life. My family is Mexican in descent. I think my grandfather came here and started living on the west side of Denver. I plan on staying here until I transfer to Metro [university] . . . [to take] early childhood education. I'd like to get into day care. Maybe open a day care. I think it's taken me a long time to figure out how to get through it. . . . Right now I'm working probably about forty [hours a week] because I'm trying to pay bills. . . . I'm going to take out a lot of loans, school loans, and just kind of do that and try to get through school, because otherwise I'll probably be in school another ten years. (Stacey, student, Community College of Denver)

Similar to Marion Bowl's United Kingdom population of adult women from disadvantaged backgrounds, who were "engaged in a long-term struggle to reach educational goals,"[16] Stacey—consistent with the parental identity of Bowl's students— is pressured by finances and time, placing her goal achievement in jeopardy.

Liz at Edmonds Community College, a returning, nonminority student in her fifties, has reshaped her life through her college educational and work experiences. But her identity as a student is far more complex than traditional concepts of college student. She is as much employee as student—as she now works on campus—and similarly, she is a volunteer worker who functions like a guardian angel for other students.

> I first showed up in 1996, [which] is when I started. . . . I was well into my age then. Nobody told me there was a difference between high school and college. . . . It was my first experience of learning what not eating and just being in the books meant. . . . I was struggling there. That was the hardest quarter I ever had. . . . And I put all my background together with the different jobs, raising four kids and as a single parent. So now, well, I have two kids finished and two boys ended up with their GED. I want to finish my one class for the . . . degree and then I was thinking of getting into medical, because I took a class more for work. It was the emergency responder: helping people in accidents. . . . I found out that I kind of really enjoyed that. So I thought, "One degree OK, why couldn't I just push it?" . . . If I wanted to go into RN [registered nursing] . . . I figure at age 52, it's a challenge. . . . So I figure I have nothing to lose and a lot to gain. But I found out if I didn't push myself, then I would've never ended up here. . . . This is not as easy as I thought. For one year, I volunteered . . . 1,200 hours helping students. I help the Hispanic students, making sure we [help] them through the first year so they don't drop [out], because that's usually when they tend to drop. So, I talk to them, see where their problems [are] and then I try to find ways that I can get them scholarships, try to find any tutors. I drag my friends in when I know that they're good at one point. [I work with] students just out of high school, and I work with a lot of the adult students that are . . . coming and looking around. . . . I am one of the security officers [at the college]. . . . During the mornings I'll do checkup on some of the students that I know are around during the mornings, and [on] my lunch hour, I kind of make my phone calls, check on some of the other students. . . . Then I have my breaks, and then there are times when I get off work early and I'll check and see if they want to have coffee or something, or run across the street and check with some of the ESL students there. (Liz, student, Edmonds Community College)

Liz, a mother of four adult children and recently remarried for the third time, has not simply expanded her personal identity but largely transformed her life's conditions, which she called "a dead-end road" living in California and in Arizona, where her family had lived "generation after generation." She moved to Washington State and arrived in Port Angeles: "I moved here with what was in the car and myself and the kids." After another marriage, she "talked the two younger boys into moving to Lynnwood with me, because I heard about the college here, Edmonds." Liz remained at Edmonds Community College as student, employee, and volunteer worker from 1996 to 2005. While she continues to pursue a variety of academic programs, with registered nursing or emergency response as her next "challenge," she has completed programs—GED, high school diploma, and an office management degree—and is employed full-time, thus she cannot be easily classified as a student. Given her lengthy connection with Edmonds Community College, both as student and employee as well as volunteer worker, her present identity is bound to the institution. Whereas Stacey at the Community College of Denver is bounded by her role as a mother who has to work to pay bills in order to go to college, Liz has an expanded assumed identity that not only encompasses the roles of Stacey but exceeds these roles because she has pushed herself. Thus aspects of one's identity can be chosen and re-created.

Multiple Motivations and Goals of Nontraditional Students

Just as precollege characteristics such as sociocultural and economic background influence students' performance in postsecondary education, and vary from group to group, so too are there motivations that influence and shape nontraditional students' decisions to participate in higher education. But motivations and goals, too, are multiple and complex. Some populations of nontraditional students tend to be motivated, on the one hand, by a desire to achieve their educational and employment goals and, on the other hand, by an enthusiasm to learn. Aslanian,[17] who examines adult students in all postsecondary educational settings, states that people in this group "lead very busy lives, juggling career and family roles. Their family incomes are higher than for most other American households. They have high levels of education as they return to college and are eager to raise those levels even further." Bay found that approximately 50 percent of adult students at one college returned to college seeking a degree in order to improve their chances of job advancement, and a significant portion of those 50 percent sought specialized job training.[18] Eppler and Harju contrasted the effects of learning and performance goals on the motivations of nontraditional students to be successful academically and persist through college to graduation.[19] They applied a psychological model developed by C. S. Dweck to nontraditional students in an attempt to understand what motivates these students to achieve. The model distinguishes between learning goals, which are characterized by a desire to increase one's competence of a skill or set of skills, and performance goals, which value outcomes rather than the process by which an outcome is reached. Older returning students tend to be motivated by learning goals and are more likely to persist in the face of obstacles than are younger students—in addition, the older the student, the stronger the commitment to learning goals. Although this trend existed among traditional and nontraditional students, the relationship was stronger among nontraditional students. Furthermore, older nontraditional students who had a stronger commitment to learning goals also tended to have higher grade point averages.[20]

Unlike traditional students, nontraditional students are motivated by their adult life experiences and the pressures and forces of work and family. In the case of Sattie from El Camino College in California, her motivations are complex and certainly not those of an eighteen-year-old freshman. Sattie's life experiences, her social and economic status, place her outside the norm of nontraditional students. In her case, educational attainment is a different factor in her identity: She does not need the credential for a job but rather for a sense of cultural approbation.

My name is Sattie; my parents are from India and I was born in South Carolina. I used to work for the DOD [Department of Defense] at first doing logistical planning for C17 aircraft, in Dayton, Ohio. I helped with replacement parts all over the world. I met my husband who was working on his master's at the Air Force Institute of Technology in Ohio, so he was still in the Air Force. He moved to California, so I came with him here. When I got here, I got a job as a . . . I got a job right off the bat. I went into a company to help them with their logistical planning; they distributed satellites in single-family homes as well as multiple dwelling units. On top of offering DirectTV programming [their main niche], we offered ethnic programming. I was brought in to originally help with the launch of a couple of stations and how we would fulfill each geographic region. When they found out I spoke and understood Hindi, and one of the biggest channels they were launching was an Indian station out of Singapore—it was called Sony Entertainment India—I just became the director of ethnic marketing programming and all these channels were under me. Complete leap. Great job. I met some wonderful people. Unfortunately, we could not get our financing to continue, so the station jumped ship and went directly to the providers like Dish Network and DirectTV. I was really good friends with the marketing person and we tried to start our own company where we [would] launch concerts . . . without knowing the amount of money and, more importantly, culturally, how different money transactions are: We understood on a corporate level, but then when it got one-on-one, it got really difficult. Then, after that, I think I went and picked up any job I could. I was too embarrassed to go back into entertainment. I couldn't land a job as high as the one I had, so I thought I didn't want everyone to know I was a loser. So let me get a job where nobody knows me. So now I work at a warehouse bakery; they provide bread for Ralph's and Kroger, and I am the office manager there. So I guess I kinda stretched my resume, or some people think I decreased it, but I love it. There is no pressure whatsoever—fantastic people. I really regret lying to them. I wish I would have come in and said, "Look, I haven't done this but I could do this, and having owned my own business I understand all the HR laws." . . . But now I am too deep in the lies and just am going to leave things where they are. In the interim, I decided to go back to school and see what I really want to do. I never had that opportunity before. Of course, my parents wanted me to be an engineer or a doctor. I thought that I should take something since it has been ten years, to see if I could get used to going back. (Sattie, student, Health Information Management, El Camino College, California)

Sattie is taking a noncredit certificate program, intending to complete a baccalaureate degree that has eluded her.

I definitely want to complete it [baccalaureate degree]. Everybody assumes I graduated. I never get past the, "no, I didn't." Suddenly I am dumped . . . whoooo, and I am the director of marketing and I am pitching Sony Entertainment. If you think about it, it is ridiculous. I think people make you what they want you to be, and I have always played that role. I am at this place that I want it for myself. It is more about me. I gotta get myself together. At this point, I feel so inept.

It is clear that Sattie has both the background to read sophisticated literature and accomplish more than a noncredit certificate program in medical billing, which is her El Camino College program. She came near to completing a business major at the University of South Carolina, and she has worked at managerial jobs. Sattie is a first-generation student with an immigrant family; her educational attainment—university course completion and near baccalaureate attainment— conforms to the norm, especially given her apparent socioeconomic background.[21] Indeed, for children of immigrants, socioeconomic status is the key factor in educational attainment.[22] However, in Sattie's case, her lack of degree attainment has more to do with her personal interests or motivation than her abilities or skills.

Not all categories of nontraditional students possess high levels of motivation to attend college, nor are they driven by clear and compelling goals. While the scholarly literature focuses upon either the tenacity of the nontraditional student or the innumerable barriers faced by this population—implying student achievement just for overcoming these obstacles— there are few accounts of students who are either overwhelmed with or out of their depth in college. Some scholars, such as Penelope Herideen and Marion Bowl, suggest that not all nontraditional students—particularly those who are considerably disadvantaged— have the capability to persist in college or to benefit from their college experiences or credentials.

Tereaza, a General Education Development (GED) student at Truman College[23] in Chicago, has several goals, but none that are clearly mapped out, and there is no expressed motivation that could ensure program completion or impel her toward a career. Her obligations and distractions do not further her college work. Even her first noted career goal—pharmacy technician—is not a certain goal because she contemplates real estate work as well. She is aware that her time and attention to study are in conflict with her family obligations.

I am from Yugoslavia. I am here nineteen years. . . . I didn't finish high school in my country and I feel like I need it. . . . I have a lot of odd jobs, like cleaning jobs. Now I feel like I have kids, and I feel like they need my help and I should be able to help them. . . . My first goal is to get GED. Then after that, I don't know, maybe I will get some degree. It takes me a long, long time because I don't feel like I get a lot of time to study at home. . . . I use to come four days a week and I have to cut it to one day a week, because the kids will be home for the summer and I don't have anybody to watch them. I feel like I can only squeeze out one day. . . . I have been here for, like, one year trying for GED. . . . I saw a program, in a pharmacy: to be a helper; that is what I would like to do. I am thirty-eight years old right now, but I would like to be a helper. That is what I would like to do. . . . I haven't [talked with anyone about that] . . . because everybody says, first you have to do GED. Finish that and then you can go. . . . I was thinking maybe real estate [too]. I like working with people, I am friendly; somehow people they like you. I was thinking maybe something like that. . . . I should study [harder] for myself. Something that would give me more pressure to study. OK, you have got to read all of this and tell me tomorrow. (Tereaza, student, Harry Truman College)

No doubt, there are conflicting pressures upon Tereaza, and her decision making is apparently a solo effort. Her career goals are tenuous and she has limited her educational goals to the attainment of a GED.

Even returning students with university degrees are not all as highly motivated or goal oriented as the literature suggests. Stanley, a fifty-seven-year-old at Wake Technical Community College[24] in Raleigh, North Carolina, lost his job and now faces a new occupational track. His goals are less than certain and his career motivational level is not high.

Our manufacturing went to China and to Mexico because [of] NAFTA . . . and because of that we closed down our manufacturing operations in Raleigh and shipped them out. Then NAFTA kicks in and says, "OK, we'll retrain you," 'cause they know we are not going to get a job. At my age you may not get a job anywhere. . . . I'll be fifty-seven this year, uh, I worked, I was four years in the military, the Air Force; I spent thirty-one years with a company. . . . I have one son and a wife I have been married to for thirty-four years. . . . I live in the country [on] six acres with a house out in the middle of nowhere. I could have gotten a job if wanted to move from the States. I had no desire to move. I could have [gone] to Texas one time and/or Mexico. I didn't want to do that. That's just not [for me]. . . . I was just too old right now; I just don't want to do that. Uh, I guess I could have worked another five to seven years, but I'll find something in this area. I am seeking a . . . what's called a[n] industrial systems technology degree, an associate's degree with emphasis on electrical wiring, air-conditioner, heating, plumbing. . . . "When I grow up, what I want to do?" I will probably . . .

I don't really know. Whether I want to work full-time or part-time, I really don't know yet. I do know that I have learned the proper way of doing, you know, the plumbing and the heating. . . . If I don't do my own business, I will try to get a job with the federal government . . . or some city municipal type work and probably in the maintenance world. (Stanley, student, Wake Technical Community College)

Stanley vacillates over his goals. Although his academic goal of an associate's degree is specific, his work or career path after a credential is less than clear. His seemingly rhetorical question—"When I grow up, what I want to do?"—leads to an answer that indicates little certainty—"I will probably . . . I don't really know"—and contrasts with the scholarship on adult students. Stanley's participation in college is not directly connected to employment goals. Instead, he was attracted to the program for its own sake, because the coursework appealed to him. "I've had a blast here. I've learned more than I ever thought I would: it's been very rewarding to me," he says. The learning experiences in and of themselves have motivated him to persist: "I've had a great time. Haven't missed a day of school; haven't missed a day of class; haven't missed even a minute of class. And, uh, it's just been really, really, really good for me."

Distinctions in student motivations and goals exist not only across age categories but also along ethnic lines, according to research investigations. African American and Hispanic nontraditional students are more likely than white students to list among their motivations to attend college the desire to set an example for others and be the first in their families to complete college.[25] The factor that most likely motivates Hispanic nontraditional students is a desire to acquire some knowledge, followed by achieving self-development, ensuring job advancement, establishing social status, and finally improving social position.[26]

Indeed, the research literature differentiates among academic goals of nontraditional community college students based on their racial classification. Common to all nontraditional student groups is the goal of obtaining a credential or degree to secure a good or better job. Among community college students in one research investigation, white students intended to obtain an associate's degree and then transfer to a four-year institution. White students also expressed an interest in pursuing a master's or doctoral degree. Hispanic students also expressed interest in following the community college experience with the pursuit of a bachelor and master's degree. African American students, however, set their sights on earning a certificate or associate's degree, and few of these students articulated aspirations of earning a bachelor's or advanced degree.[27]

Combined with the variable of race or ethnicity, nontraditional students who are women exhibit motivations that differ from other populations. Veronica Thomas found that the main goals of women reentry college students were personal fulfillment—including enhanced self-image, self-empowerment, and higher esteem—and opportunities for job placement, as well as career development.[28] These goals were framed within a context of benefits to their children. Latrenda, an African American adult student at Community College of Denver, works, attends college, and is motivated by her exemplary role for her children: "[I am] working, going to school, and raising three teenagers. It is hard; it's not an easy task. My, I guess how I do it is staying focused, setting my goals actually and achieving those goals, and by [setting] my goals, my children also set goals for themselves and try to achieve those. So I . . . set a good example for them." Latrenda moved from completing a GED to an associate's degree in applied computer science. Although nontraditional African American women, particularly women with children, are motivated by several factors different from those of other groups, nontraditional students as a whole experience difficulty with balancing responsibilities associated with their multiple roles. W. Norton Grubb and Marvin Lazerson refer to this as a "work-family-schooling dilemma."[29] Kathleen Shaw found that community college students often feel fragmented and forced to choose among and balance their multiple responsibilities.[30] With the majority of community college students enrolled part-time, primarily because these students have to work, their motivations to persist in college must be considerable. As older students, they often have commitments and responsibilities to spouses and/or children, and thus their focus upon college work requires tenacity or else the academic outcomes will be far from satisfactory.[31]

Unquestionably, characteristics of nontraditional students are influential in their motivations and educational goals. Haleman argues that for many single mothers, who are a significant component of nontraditional students, the choice to pursue higher education is a mechanism for moving from poverty to the middle class.[32]

> Their educational experiences often provide an opportunity for disrupting negative expectations directed toward them and also symbolize for the women the possibility of realizing the hopes and dreams they hold for themselves and their children.[33]

Haleman notes that for these women, higher education is an escape route from poverty. Her study looks at single mothers who often remain in poverty even after working full-time, and she compares this condition to that for males, who generally are only in poverty when unemployed.

> Single mothers are thus especially needful of postsecondary education and the increased earning potential it represents in the present context. The economic difficulties experienced by many single mothers are compounded by a lack of education that virtually guarantees the continuing poverty of single-mother-headed families.[34]

Furthermore, their roles as students become an important part of their identity. Haleman also suggests that the decision to pursue higher education is not just a choice that single mothers make for themselves; it contributes to the "educational pipeline" for their children, too. Thus, their participation in and success within postsecondary education add to the social capital that they can then pass on to their children as they matriculate through their primary and secondary education.

Tina at Virginia Highlands Community College expresses a sentiment common to adult women with children. She indicates that her motivation was enhanced by her teacher.

> Well, I guess if I had it to do over, I'd started right after high school. I'm thirty-five years old now, and I have two children—six [and] four—both little girls. Divorced. Kara and Kelly. And you know, I kind of decided I didn't want to be in a factory for the rest of my life. So I decided to go back to school. I took Toby's CNA [nursing assistant] class. And she's wonderful, of course, so I'd give anything if she taught nursing classes, especially now that I got in. So I learned a lot from her, basically, and I learned that who I want to be when I grow up is Toby.

Tina's emphasis upon her children suggests that they both contribute to her motivation but also structure her work as a student.

> But I'm able to take this summer off. My kids will be home, it'll kind of be hard for me to go to school full-time in the summer, cram up the course, and then I'm going to try to rest up, get everything together so I can start and just go at it full force. Hopefully, by the time I'm done, I'll have a couple of degrees, and once I start my nursing job, still going to try to get my master's in human services and counseling. . . . So I have two goals, and I'm sure one I'll probably reach before I reach the other one. And I've got my kids involved. They know what I'm doing. Of course, they went out to eat when I found out, when I got [accepted to the program]. I was excited and the girls—in fact the way they acted helped me a lot because I know they're excited like I am. My six-year-old is, like, "My momma's going to be a nurse; she's going to help people get better." She brags on me. . . . And I feel like if they see me trying to do better for myself then . . . it'll give them the gumption to try to do everything. Hopefully they'll go to college after high school and not be like me and just get out and work. . . . But there are some days that they just want me to be home with them. And that's why this time I tried to get everything during the day while they were in school. I've got my schedule where I can put them on the bus and get them off the bus in the evening; got my classes compacted in between. Because I don't want them to have to suffer for me doing what I want to do. . . . But they're excited for me, and every time I look at them I know why I'm there: because some days I have thought about quitting and going to work. Just because money is tight, things get hard, really; got a family to raise; got to have a job. . . . When you're older you appreciate the value of getting an education and having a good job. . . . Every time I look at my girls, that's why I'm here—for me and them. . . . Probably if I didn't have them, I would probably be less apt to do as well as I've been doing. (Tina, student, Virginia Highlands Community College)

The working poor or low-wage workers who attend college are another distinct population, although they are usually members of other identified populations, such as minorities, whose goals and motivations are affected by their experiences and circumstances.[35] This population is subject to a host of impediments and barriers—including, of course, financial barriers—but these students face additional stresses as well, such as balancing their work, family, and college life. Furthermore, their financial needs are not assuaged if they attend college on less than a half-time basis, as this disenfranchises them from several forms of financial aid.

The decision for a low-income student to enroll in college, and with the intention to persist to completion, revolves around the question of how to pay. Expenses for college include not only tuition and fees, but also books and related materials and living expenses such as food, housing, and transportation. Ironically, need-based aid is lowest for community college students because tuition costs at community colleges are lower than at other post secondary institutions; yet, larger numbers of low-income students apply to community colleges than other institutions. As reported in 2000, the average grant for community college students without dependents was $2,200, while the average grant for dependent students at private four-year colleges was $7,900.[36] The pressures of having to work while in school, worrying about unmet needs and covering bills, possibly combined with other family responsibilities, are significant. These pressures are correlated not only to high-risk factors for students in their educational attainment, but also to conditions that shape college experience.

A subset of the working poor includes welfare recipients who attend college. James Valadez interviewed African American women living in a rural town to learn about their experiences in the welfare-to-work program at the local community college.[37] All were single mothers, some were high school dropouts, some were workers who were displaced after a large factory in town closed, and some had never worked. The program was structured so that the women would meet four days a week for ten weeks, and it focused on professional appearance, personal discipline, interviewing skills, and positive work-seeking attitudes and behaviors. When Valadez followed up with the program graduates, some of them had been unable to find work because of lack of transportation or lack of opportunities in the area. Some had found minimum-wage positions, but they quit within a few months because the income could not sustain their families. Valadez reflected on the barriers of race, gender, and class bias, and the difference between encouraging a student to have a new ideology about work compared to the reality of constraints in their environment. These women possessed student identities by fiat: The conditions of their role were set out by federal policy and state/regional application of policy. In this case, the outcomes of educational experiences simply reinforced their low-income identities.

Although a large percentage of nontraditional students have goals and motivations to pursue postsecondary education for credentials, only half of adult students, a subcategory of nontraditional students, attend colleges for credentials, while the other half do not.[38] This suggests that there is a large population of nontraditional students who are engaged in postsecondary education outside of the for-credit system, through continuing education programs and industry training. While this pattern of behavior is somewhat outside the discussion of this book, an aspect of the pattern is addressed in Chapter 7 (Continuing Education and Lifelong Learning).

Educational Goals of Nontraditional Students

While the educational goals of nontraditional students are varied, they tend to be specific, such as the attainment of a skill such as English language acquisition, a degree and credentials for employment, and the completion of a course of study previously pursued. Other goals, such as career goals and life goals, are furthered by participation in college study. Ellen at Bakersfield College in California expresses both her educational and career goals, including the rationale for these goals. Ellen is an African American woman in a predominantly white and Hispanic community.

> My name is Ellen. I am forty-one years old. I have four children. I have two boys in college, ages twenty-two and nineteen. I have a twelve-year-old daughter in junior high and I have a two-year-old. I have been here at Bakersfield College since . . . 2000. I have had a lot of setbacks due to health problems. I had several surgeries. Of course, my pregnancy was considered high risk, so I had to take the whole year off for that. Other than that, I am OK. You know, life happens, and I have had a lot of things go on. I am a human services major. I plan to transfer to Cal State University after I am done here. I wanted to do something different . . . by helping people get off drugs, and other people do this, but when I came in and started seeing how many people who are . . . on drugs, on alcohol and stuff, but they wanted to change their lives. Well, of course, you have to get that out of the way, but sometimes if you just give them something new to focus on, they don't want the drugs anymore, and I saw that firsthand. And I think once you gain their trust and get them to open up and realize that they are more than they have been told they are, they will open up to other options, but a lot of people just . . . well, you need to go get help; you need to go to rehab. "Well, after rehab, what am I going to do? I have been getting high my whole life. I don't know how to do anything else." So, you have to kind of put those other options in front of them, this is my opinion, in order to let them know that once I am clean, I am going to do that. Then it is easier. If you just want them to get them get clean, OK. "I am clean, now what?" They are going to go back, because they are going to go to the familiar things. There is more out there, there is more things that I can do with that [bachelor's degree]. I do know a long time ago, the AA degree was like a really big thing. You can do a lot; it is almost a high school diploma now. And I am, like, OK, but it is important. I am not downplaying it in any way. It is very important to have it, but you know, I

think now, you are going to have to take that extra step; you are going to have to keep up with the times. (Ellen, student, Bakersfield College)

Harold articulates his presence in an adult high school completion program at Wake Technical Community College in North Carolina, in the Basic Skills Center.

I am nineteen and a half years old, almost twenty. This is my fifth year in high school. I came here because I had difficulties at Athens High School, getting into the wrong crowd. I was getting tired of the whole teacher situation thing. My parents told me that Wake Tech offered a high school diploma program. I wanted to get a high school diploma. It was either this or go back to Athens. I found out I could qualify for this. I didn't want to get a GED even if they were equivalent. I really wanted a high school diploma. (Harold, student, Wake Technical Community College)

Harold's educational goals are specific although limited—a high school diploma. Rosa at GateWay Community College is undertaking a chef's apprenticeship program to prepare her for her ultimate educational goals—attendance at an elite chef's school: "I hope I can go to [Le] Cordon Bleu. I know it is one of the expensive schools, but I know it is one of the best." She is at GateWay because she is unable to afford the costs of more prestigious schools. She is in the process of making her way to a career as a chef, and thus she is accomplishing learning goals.

Well, the chef's apprentice is . . . first of all, you learn to cook. It tells you how to make stuff; why [it] has to look nice, why [it has] to not . . . be burned, why [it has] to have a nice taste, [and] everything you're going to use. [The program] has a lot of things, very interesting things, especially if you like to cook. So, I love to cook. Everything I make every day looks interesting to me. (Rosa, student, GateWay Community College)

Of course, Rosa's first learning goal to accomplish was English language skills, since she is from Mexico and her first language is Spanish.

Adrene, who is twenty-six and has been at multiple higher education institutions as well as in the workforce, has found that her learning experiences have led to the development of her educational goals.

I've learned about how I can touch people, which I'm kind of surprised about. I did a speech in my speech class and I had somebody come up to me and give me a hug and start crying afterward and I was so, like, "What the heck is going on?" But I guess I was talking about the things that I'm involved in, and I talked about the women's forum and I talked about student ambassadors and all that, and she was actually the founder of the women's forum and so it touched her to have somebody actually speak on it that wasn't originally a part of it, [so] that she could see what her work had done. And so that's been pretty cool. And they get me involved in a lot of community service stuff, so I get to learn about different people too, because being here I'm around so many different kinds of people. We have a high school here also. And so, some of the students in my classes have come through the high schools because if you go to GateWay High School, then they'll pay for your classes here at GateWay Community College. And so it's interesting to see the diverse group of people; [there is] one guy who's just got here from Africa, then there's the kid that just came from high school upstairs that's in the class. And so I get a lot of perspectives and different viewpoints. (Adrene, student, GateWay Community College)

Even though Adrene was a strong high school student, she did not adapt well to her first year of college—at a state university—just out of high school. Her eventual return to college and subsequent enrollment at GateWay led her to pursue nursing.

When I first came here I wanted to become a registered nurse, so I took some . . . nursing classes that get you started, like fundamentals in health care and stuff like that, and I liked it. It was just the idea of being a nurse, after I heard people say, well, you can identify infections by smell when you walk in a room. And I thought, well, I don't want to smell nothing and know what it is. I don't want to be, like, "Oh, that's a staph." You know? So, that wasn't for me.

Next came elementary education. But then she moved toward psychology and further education aspirations, which included transferring to a university and then moving on to graduate school.

OK, elementary education will work for me. OK, this will work for me. Finally . . . when I first started going to college, psychology was what I wanted to do, and so I've taken, what, three out of the four psychology classes that are offered here, and I still like it. So I decided that's what I would do . . . get my AA in psychology and transfer over to ASU [Arizona State University]. So I am going to transfer back to the big school anyway, but transfer to ASU to work on getting something in child psychology: my bachelor's, maybe a master's, if I can afford it and got the time. (Adrene, student, GateWay Community College)

Educational goals are focused on both learning goals—such as language acquisition—and on performance goals, such as a high school diploma or degree attainment. For nontraditional students, learning goals and the process of fulfilling them develop performance goals. While there is a considerable body of research that claims student educational goals, such as program and degree completion, are major factors in student persistence, for nontraditional students, especially older students such as Adrene, the behaviors and goal setting are neither static nor singular. The accomplishment of learning goals along the path of college attendance can result in the alteration of performance goals. Furthermore, noncurricular and other experiences outside of the classroom can lead to the fulfillment of learning goals.

The Student-Institution Disconnection

As early as the 1980s, scholars identified both tensions and disarticulations between student populations and practitioners at community colleges.[39] With enrollment growth and the realization of access for diverse populations there is considerable potential for misalignment of institutional functioning and student needs. In addressing or solving one problem, institutions may initiate and sustain new problems. To ensure higher retention rates, colleges turn to higher admission requirements for some programs, resulting in a large student population in general studies programs or other programs that are not part of the educational goals of students. Thus, student motivation drops, as does student performance, because students find themselves in the wrong program, one that does not fit their goals.[40] In North Carolina, a high school credential or equivalent is required at community colleges for college-level programs, and those who do not have this credential must undertake remedial courses work. The state does not fund remedial coursework at the same level as college-level coursework, and remedial students can be and are separated both physically and educationally from mainstream college students. Furthermore, because of the funding pressures, noncollege-level courses are staffed largely by part-time faculty whose academic credentials are less than those at the college level.

> Because we are still accredited with the association with colleges and universities, [faculty] have to have at least a baccalaureate degree. And we like them to have at least two years of experience teaching. Some don't because that isn't always possible. We also have a preservice training that we put into place, ten to twelve hours, including an apprenticeship or student teaching. . . . Then they must have at least twelve hours of in-service training throughout the year. Then we employ them. (Dean, Adult Basic Education, Wake Technical Community College)

Thus, students with the least academic abilities and the most social and economic challenges, such as new immigrants, students with disabilities, and those who require basic academic credentials (e.g., GED) to allow admissions into credit-bearing programs, are subject to an education that relies upon lesser resources than those students who are more able and fortunate in social and economic terms. This is not to suggest that community college students as a whole are fortunate socially or economically, but rather that by privileging one group, the institution has further disadvantaged a large population.

Social class and ethnicity are other dividing lines between students and the institution. For the large minority student population at Mountain View College, their struggle is with an unfamiliar environment, where faculty members are not ethnically diverse, and with learning requirements that do not reflect their previous learning experiences. They may have passed the Texas high school graduation test, but they do not adjust to the higher expectation of college-level work. While they are not as disenfranchised as students who do not possess high school diplomas at such colleges as Johnston Community College in North Carolina, or Pima College in Arizona, or Edmonds Community College in Washington, their college performance may lead them to drop out, and that circumstance will soon place them in a similar category as they will qualify only for service jobs, such as custodial work or manual labor.

For the vast majority of students at Mountain View, the problem is not access to classes but access to further education as student performance impedes progress. Students of varied backgrounds and abilities are subject to the decision of the college to provide low-cost education for the institution, what the college dean refers to as the "cash cows."

> We have a retention problem here. . . . One of the things that may even exacerbate it is the use of technology such as online classes. . . . Sometimes as much as 80% of the students drop or don't complete. . . . It's a difficult situation. . . . It's always the online classes that have the highest level of . . . the highest problem with rates of retention. . . . The online classes are the cash cow. Students come in late [to register, but they] can take online classes. (College Dean)

Technology in the form of online classes actually hampers student retention and creates an unfair condition for nontraditional students.

Indeed, the use of technology cuts two ways. First, institutions, in order to expand access and cope with diminishing state funding, resort to online and distance education, an approach that favors specific populations, such as those who are technologically experienced and those who can learn without face-to-face interaction.[41] Second, students who are classed as highly nontraditional or "beyond the margins" have little access to the technology that institutions provide for other students, and their coursework and programs make little or no use of technology in instruction. Students in Adult Basic Education or basic skills programs, including English as a second language (ESL) programs at such institutions as Truman College and Wake Technical Community College, indicate that electronic technology, excluding the use of computers for drill-type activities, is not part of their college experience. Thus, technology has the potential to stratify student populations in the same way that it perpetuates divisions outside of postsecondary education.[42]

With respect to these institutional and student misalignments, college officials do recognize the gaps that exist between the promise of postsecondary education and the condition of students.

> We actually [take] students to four-year institutions so that they can see the process of the admissions and so that they can tour and that kind of

thing, because they don't have the money to do it themselves or maybe they don't have the motivation or don't know how to do it. [W]ith first generation students . . . they don't know what a major is, or what we call it here: We call it a program of study. And so, when you use the term *major*, you confuse them, and what's a minor and what's a concentration? And just the terminology is so foreign to them they're intimidated whenever they talk with counselors from four-year schools, so we . . . help them make that transition, and it seems to be really useful. And they seem to enjoy that. And so with the workshops and the four-year college visits, we usually run about two a month and students can choose to come. And then we also run a huge tutorial program. And that is free to all the students in our program. There are only 160 in our program. So if they want to receive tutoring, and specifically math, science, and English, then they can, but we also do some in accounting or word processing if there are still students [who have] needs, in order to help them in those basic classes. . . . On our campus there are also returning students, returning adults. . . . The average age in my program is thirty-four. So they're generally an older population. We have one participant that's seventy-two years and we have . . . several participants that are eighteen years old, so it's a very big range, which makes programming a little bit difficult for them. . . . The majority of them are coming in off of layoffs. . . . The tutoring that we offer, right, for the students, obviously there is a much greater need for that than what we have, but with four people on staff that's all that we can maintain. (Jennifer, Counselor, Johnston Community College)

In describing college services for students, Jennifer also acknowledges the considerable unmet need of students. With only 160 students in the program because of college resources, Jennifer implies that many other students who need these services are shut out. Nonetheless, the 160 students are socialized and educated for the culture of postsecondary education.

At the Community College of Denver, where access to postsecondary education is a mantra not only for college identity but also for college practices, Alisa, a college dean, explains the lengths to which the institution will go to provide access to the most marginal members of the community. The issue of undocumented immigrants is not brushed off; rather, it is addressed in ways that allow the college to interpret the applicable laws and regulations.

This college really has struggled with the notion of undocumented students, and we have for the longest time been the only institution in Colorado that would accept undocumented students and everybody . . . said, "How does CCD do it? Why do they do it?" It's because we believed it's important and the right thing to do. Since 9/11, we have had to change a little bit. . . . We interpreted the law, the financial aid. We interpreted the regulations for establishing residency as having established a domicile, having established physical presence in the state of Colorado for 365 consecutive days. That's how we determine residency. Some institutions, now, [say you are not a resident] if you don't have a Social Security number, and lots of people don't remember their Social Security numbers, and so . . . if you don't have a Social Security number we assign you an ID number and we had never asked [for] more information than that. There's a community college in Colorado that . . . will require you to write a letter to the admissions office stating that you're an undocumented citizen. I don't know why anybody would do that. I'm astounded that that would be a requirement. Now if you're a Denver public schoolkid, we make the assumption that you're a resident for the purposes of tuition—or if you're a public schoolkid in Colorado. . . . We will accept anybody. (Alisa, Dean, Community College of Denver)

While there are no doubt numerous other problems and challenges for undocumented immigrants, their educational goals at least had an opportunity to come to fruition at Community College of Denver, because the institution took the first step in admitting this population. However, by November 2005, the Community College of Denver, along with other Colorado postsecondary institutions, was legally obliged to require undocumented immigrants to pay out-of-state tuition, which was set at five times in-state tuition. Although this state policy is contested, the Colorado Commission on Higher Education held to its policy as recently as March 2007.[43]

Beyond the organization of individual community colleges are large social problems and arguably failures—failures of the state and its institutions generally. A large population of potential students are disadvantaged but not served by community colleges. They do not have educational experiences at the postsecondary level to compare with disadvantaged students whose experiences suggest triumph over adversity. Large numbers of eighteen- to twenty-four-year-olds who have left high school are untouched by the community college, as noted by the chancellor for California's community colleges.

[In 2001], we had over a million young people between age 18 and 24 who did not possess a high school diploma. Now, today that is probably 1.2 million . . . [and] the youth unemployment rate, that is, the 18 to 24 unemployment rate, is about 22%. . . . Some of them are washing cars; a lot of them are just wandering the streets and being street dwellers. I mean they are out there and they are clogging the health care clinics when they get hurt; they go into the emergency rooms; they are on any form of public assistance that they can find. I mean they are not paying their own way. They are not paying taxes; they are not helping the school systems, yet they are having kids. . . . It is a huge problem. . . . Frankly, today we are not structured, nor are we funded, to address that population. We don't really address them. Some of them wander in; we assessment test them, [and] if they test below the tenth grade they have to go to noncredit/remedial stuff. . . . I mean you can't do anything. We know how—I mean there is research on how to help these people overcome their skill deficiencies—but it is expensive. It takes small classes; it takes six hour long classes; it takes daily classes, all kinds of attention. . . . So we just don't do it. We do give these classes, but we pack eighty people into a class, give them a tape and a headset and say sit there and listen to the stuff and tell me how it was. I mean, it is miserable. . . . That is a miserable failure; so you don't get it at our place you get it over at adult ed, [which] is even worse. I mean mortifying, completely mortifying. . . . I would say if there is 1.2 million of them, we probably touch, maybe, we see 50,000 of them a year in a cursory way. They come in because they kind of heard there may be something there for them, but there is really nothing there for them, the way we are set up. Adult ed really suites their need better than we do, but high school already failed these people. They are not going to go back to high school and go into adult ed. So, they are just out there, bitter, screwed, and out of it.

Unlike students who flourish in community colleges, this disenfranchised population is characterized by the chancellor as the "miserable failure" of the community college.

While the community college, among all postsecondary educational institutions, offers the most benefits to nontraditional students, particularly those who are highly disadvantaged, and has functioned to advance nontraditional students who would otherwise not attend postsecondary education,[44] there continue to be large gaps between the potential or promise of fairness for disadvantaged populations in comparison to those with advantage. Certainly one reason for the disconnection is the pressure to respond to the demands of government, business, and industry to meet global economic and workforce needs;[45] this pressure has led the community college to compromise its core values. A second reason is that there is simply not enough money to serve everyone, and the policies and funding behaviors of state and federal governments do not favor the disadvantaged student. Certainly another reason is that, although the community college endeavors to meet the needs of the disadvantaged, an educational institution alone cannot fulfill unmet needs of this magnitude; other institutions, such as health care, labor, and social services, have a responsibility as well.

Notes

1. Benedict Anderson, *Imagined Communities: Reflections on the Origin and Spread of Nationalism* (New York: Verso, 1991); Dorothy C. Holland et al., *Identity and Agency in Cultural Worlds* (Cambridge, MA: Harvard University Press, 1998).
2. Holland et al., *Identity and Agency in Cultural Worlds*.
3. Ibid.
4. Anna Sfard and Anna Prusak, "Telling Identities: In Search of an Analytic Tool for Investigating Learning as a Culturally Shaped Activity," *Educational Researcher* 34, no. 4 (2005): 14–22.

5. Rebekah Nathan, *My Freshman Year: What a Professor Learned by Becoming a Student* (Ithaca, NY: Cornell University Press, 2005).

6. Tom Wolfe, *I Am Charlotte Simmons* (New York: Farrar, Straus and Giroux, 2004).

7. Marion Bowl, *Nontraditional Entrants to Higher Education* (Stoke on Trent, UK: Trentham Books, 2003); John E. Roueche and Suanne D. Roueche, *High Stakes, High Performance: Making Remedial Education Work* (Washington, DC: Community College Press, 1999).

8. Kathleen Shaw, "Defining the Self: Construction of Identity in Community College Students," in *Community Colleges as Cultural Texts*, ed. Shaw, James Valadez, and Robert Rhoads, 153–71 (Buffalo: State University of New York Press, 1999).

9. Alicia C. Dowd and Randi Korn, "Students as Cultural Workers and the Measurement of Cultural Effort" (paper presented at the annual meeting of the Council for the Study of Community Colleges, Boston, MA, April 8, 2005).

10. Kenneth Gonzalez, Carla Stoner, and Jennifer Jovel, "Examining Opportunities for Latinas in Higher Education: Toward a College Opportunity Framework" (paper presented at the annual meeting of the Association for the Study of Higher Education, Richmond, VA, 2001), 2–37.

11. Laura W. Perna, "Differences in the Decision to Attend College among African Americans, Hispanics, and Whites," *The Journal of Higher Education* 71 (2000): 117–41.

12. Patricia McDonough, *Choosing Colleges: How Social Class and Schools Structure Opportunity* (Albany: State University of New York Press, 1997); James Valadez, "Cultural Capital and Its Impact on the Aspirations of Nontraditional Community College Students," *Community College Review* 21, no. 3 (1996): 30–44.

13. Ricardo D. Stanton-Salazar, "A Social Capital Framework for Understanding the Socialization of Racial Minority Children and Youth," *Harvard Educational Review* 67 (1997): 1–29.

14. Minerva Santos, "The Motivations of First-Semester Hispanic Two-Year College Students," *Community College Review* 32, no. 3 (2004): 18–34; Watson Scott Swail et al., "Latino Students and the Educational Pipeline: A Three-Part Series" (Washington, DC: Educational Policy Institute, 2005).

15. Patrick T. Terenzini et al., "First-generation college students: Characteristics, experiences, and cognitive development," *Research in Higher Education* 37, no. 1 (1996): 1–22.

16. Bowl, *Nontraditional Entrants*.

17. Carol Aslanian, "You're Never Too Old: Excerpts from Adult Students Today," *Community College Journal* 71, no. 5 (2001): 56–58.

18. Libby Bay, "Twists, Turns, and Returns: Returning Adult Students," *Teaching English in the Two-Year College* 26, no. 3 (1999): 305–12.

19. Marion A. Eppler and Beverly L. Harju, "Achievement Motivation Goals in Relation to Academic Performance in Traditional and Nontraditional College Students," *Research in Higher Education* 38, no. 5 (1997): 557–73.

20. Ibid.

21. Katalin Szelényi and June C. Chang, "Educating Immigrants: The Community College Role," *Community College Review* 30, no. 2 (2002): 55–73.

22. Jennifer E. Glick and Michael J. White, "Postsecondary School Participation of Immigrant and Native Youth: The Role of Familial Resources and Educational Expectations," *Social Science Research* 33 (2004): 272–99.

23. Personal names of interviewees at Truman College are pseudonyms, because of an agreement with the institution, for purposes of collecting data.

24. Some of the students at Wake Technical Community College are given pseudonyms because of an agreement with the institution.

25. Julie Weissman, Carole Bulakowski, and Marci Jumisko, "A Study of White, Black, and Hispanic Students' Transition to a Community College," *Community College Review* 26, no. 2 (1998): 19–42.

26. Santos, "Motivations."

27. Weissman, Bulakowski, and Jumisko, "A Study of White, Black, and Hispanic Students' Transition."

28. Veronica G. Thomas, "Educational Experiences and Transitions of Reentry College Women: Special Consideration for African American Female Students," *Journal of Negro Education* 70, no. 3 (2001): 139–55.

29. W. Norton Grubb and Marvin Lazerson, *The Education Gospel* (Cambridge, MA: Harvard University Press, 2004).

30. Shaw, "Defining the Self: Constructions of Identity in Community College Students," in *Community Colleges as Cultural Texts*, ed. Shaw, Valadez, and Rhoads, 153–71 (Buffalo: State University of New York Press, 1999).

31. William Maxwell et al., "Community and Diversity in Urban Community Colleges: Coursetaking among Entering Students," *Community College Review* 30, no. 4 (2003): 21–46.

32. Diana Haleman, "Great Expectations: Single Mothers in Higher Education," *International Journal of Qualitative Studies in Education* 17, no. 6 (2004): 769–84.

33. Ibid., 770.

34. Ibid., 775–76.

35. Lisa Matus-Grossman and Susan Gooden, *Opening Doors: Students' Perspectives on Juggling Work, Family, and College* (New York: Manpower Demonstration Research, 2002).

36. Susan Choy and Larry Bobbitt, *Low-Income Students: Who They Are and How They Pay for Their Education* (Washington, DC: National Center for Education Statistics, 2000).

37. James Valadez, "Searching for a Path out of Poverty: Exploring the Achievement Ideology of a Rural Community College," *Adult Education Quarterly* 50, no. 3 (2000): 212–30.

38. Lisa Hudson, "Demographic Attainment Trends in Postsecondary Education," in *The Knowledge Economy and Postsecondary Education*, ed. P. A. Graham and N. Stacey, 13–54 (Washington, DC: National Academy Press, 2002).

39. Richard Richardson, Elizabeth Fisk, and Morris Okun, *Literacy in the Open-Access College* (San Francisco: Jossey-Bass, 1983); Lois Weis, *Between Two Worlds: Black Students in an Urban Community College* (Boston: Routledge and Kegan Paul, 1985).

40. Grubb and Lazerson, *The Education Gospel*.

41. John Levin, Susan Kater, and Richard Wagoner, *Community College Faculty: At Work in the New Economy* (New York: Palgrave Macmillan, 2006).

42. Vicki Smith, *Crossing the Great Divide: Worker Risk and Opportunity in the New Economy* (New York: Cornell University Press, 2001).

43. This policy was evident in the meeting minutes of the Colorado Commission on Higher Education and was pointed out to me by the president of Community College of Denver, Christine Johnson, in April 2006.

44. Arthur Cohen and Florence Brawer, *The American Community College* (San Francisco: Jossey-Bass, 2003); Grubb and Lazerson, *The Education Gospel*.

45. John Levin, *Globalizing the Community College: Strategies for Change in the Twenty-First Century* (New York: Palgrave, 2001).

Chapter 4

The Ways in Which Students Experience College

Students with disadvantages struggle to reach personal and educational goals.[1] They struggle because they are impeded by barriers—physical and mental challenges, family responsibilities, lack of resources, and discrimination, as well as their own lack of interest, lack of information, and a lack of guidance. They face economic, institutional, and social constraints, as well as personal limitations. In this chapter, I present these students and the condition of their nontraditionality through the perspective of students themselves as well as those who work with them. These perspectives reveal the experiences of these students in their pursuit of personal and educational goals through their attendance at a postsecondary institution. Furthermore, their struggles constitute evidence of their disadvantages and the benefits that they require in order to have their condition adjusted and receive fair treatment.[2] Such adjustments will enable individuals to flourish.

This discussion is not a series of revelations about what students learn, especially not cognitive development, in college, nor [is it] about student experiences after college that rely upon college learning. Marcia Baxter Magolda illuminates student learning during and after college in her examination of traditional age college students through narratives of stories about and by students.[3] Her work reflects something quite different from my stories: hers are testimonies of how college leads to the personal self-development of eighteen-year-old students as they mature through their twenties. Her students come of age. My students on the whole, in contrast, are of age, and their stories are about coming to college and who they are and who they become at college.

The Travails of Nontraditional Students

The discussion in this section is organized around the travails of nontraditional students. These travails include both the struggles of students who persist in college, including their life-altering experiences and their coping strategies, and those who are nonpersisters, including the misalignment between student needs and institutional programs and services, the conditions of adversity that over-whelm students, and the problems for younger students with disadvantaged backgrounds that manifest as inadequate academic skills. The section begins with those students who are among those at the margins of our society, those with both mental and physical disabilities.

My observations at Johnston Community College in North Carolina are where the narrative begins. It is May 28, 2003, and this rural community col-lege—which is located in an area about a forty-five-minute drive from North Carolina's capital, Raleigh—has its summer program underway. In describing a large group of students at his campus, the college president notes that these stu-dents "are community college students, economically disadvantaged: reentry mothers and fathers; a specialized population [that] needs extra support." Some are mentally and physical disabled students who take compensatory education programs where they learn life skills, social skills, personal hygiene, and perhaps some vocationally oriented skills. I talk with their instructor, Karen, who gives me further understanding of these students—several of whom are kept at home, out of the public light, and others who live in group homes, one even in a "rest home." They are in their late twenties or early thirties; they have few academic skills with no more than a grade-five reading level; and some have personal assistants with them at school for their health care needs. The instructor's rationale for her work with these students is altruistic, to say the least: "Don't ever give up on anybody; it's like hunting for buried treasure."

Across the hall from my interview room is the college's exercise room—a room with windows into the hallway, full ground-to-ceiling mirrors, and exer-cise machines and weights. The day after my interview with the instructor, while I am walking down the hall, I notice a group of people in this room, and then the instructor waves to me from within, beckoning me to enter. She welcomes me and gains the attention of the students: "This is Mr. Levin from North Carolina State University. He is visiting with us this week." These students pres-ent themselves as visually different from me: One student is in a wheelchair, sev-eral of them are seriously overweight, with poor posture, and some exhibit lack of physical coordination as they walk toward me and gather around me. Most reach out their hands, whether right or left, to shake my hand or touch me; many of the dozen students seek out my eyes and say, "Hello, Mr. Levin, I'm Jake, Sally, Grace." A few of them pick up weights, and one man flexes his biceps to show his instructor. These students are both awe-inspiring and heart-breaking. In this room, this place, they are at college. The wheelchair-bound student, probably thirty, says, "Mr. Levin, do you know what I'm going to do when I grow up? I'm going to be a boxer." I reply, "Don't hit me," as he clenches his fist in jest. Another male, about twenty-five, says, "I'm a student at Johnston Community College," and gives a big grin. They are white students and black students, males and females. They don't look like college students. They don't look like they will be nurses or computer analysts or office workers or assistants to teachers; not den-tists, lawyers, engineers, or business tycoons. They will have difficulty with liv-ing on their own. They will probably never be fully responsible citizens—they won't serve on juries, won't vote, and won't understand the institutions that guide, influence, and shape them. Yet, they are community college students, adults at the college to learn.

Both the institution and the instructor, Karen, enable these students to participate as students in college, whereas those who rely upon traditional conceptions of college and its expected outcomes—including, for example, that college is for advanced postsecondary learning, career preparation, and intellectual development—might reasonably exclude this population from college. Furthermore, even colleges that permit students such as Karen's to enroll do not necessarily provide the level of accommodation or treatment of these students that will allow them to flourish. Students similar to these in other colleges may be segregated, cut off from the mainstream operations and activities of the institution, because they are housed in facilities that are removed from the center of college or from the college's campus altogether.

Another form of segregation and unequal treatment begins outside the institution, at the state or federal policy level. An Adult Basic Education administrator at Pima Community College in Tucson, Arizona, characterizes the students in the college's program as a disenfranchised population. Pima College has a large number of Latino students, with 30 percent of the credit students in the category, not including these Adult Basic Education students who are classified as noncredit students and outside the category of a minority population that might receive benefits from their placement in the for-credit category.

> For the most part, our students are almost always economically disadvantaged; they are by definition educationally disadvantaged. Most of them are underemployed or employed. I mean they are working on the low end of the economic scale. Many of them are parents. These are people, in other words, on the periphery. . . . If you wanted to describe, for example, a disenfranchised political constituency or a weak political constituency, you would describe them in almost the same exact way: poor, undereducated, disengaged, working on the periphery of the economic industry, and wanting something else for themselves and their family. (Dean, Adult Basic Education, Pima College)

Furthermore, the dean notes that state policy actions essentially use the college as a solution for youth unemployment and school dropout.

> As you probably know, we have . . . I think there is an audit going on [to determine] whether or not we have the second or worst dropout [in this state] in the nation. . . . We have about a 50% dropout rate for Hispanics. The Native American dropout rate is even worse. One point I want to make, just in case we don't touch it later, is that a large number of people in the program are youth, about 25% between the ages of sixteen and twenty-three, and many of them are youth that have dropped out and are coming back for their GED or for basic skills, who have elected for whatever reason

to not return to the K through 12 mainstream. What we become, then, is a huge *de facto* youth alternative program.

In addition to the descriptions of students by college faculty and administrators, the stories of students about themselves are not only compelling narratives about personal backgrounds, struggles, and motivations, but also windows into the critical elements of persistence and achievements. Lucy is an international student at Edmonds Community College in Washington State. Students who arrive in the United States with an F1 visa for the sole purpose of studying are considered international students, as differentiated from adults who come with a work or tourist visa.[4] International students must maintain full-time course load and abide by many regulations to maintain their F1 visa. Although Lucy is officially an international student, her plan is to remain in the United States, and thus she views herself as an immigrant.

I'm in my forties and I'm from Singapore. . . . I have been here for four years. . . . I worked in Singapore in the hotel. I worked in the Four Seasons hotel and I always wanted to further my education, but during my time, especially my time I was there, it was very uncommon to have a woman have a high education. I was lucky. I had an education, although not a very high education. I finished high school. . . . During that time everything is focused on the boys, men. They had the right to further education but not the women. Women are supposed to get married, be wives, and mothers. . . . I want to have a further education. Even though my parents could afford it, they said, "No. . . . Your brothers go to school, you don't need it because . . . you are luckier than most women, because most women just finish elementary school four or five years and that's enough. Because that's enough, you don't need the further education. What are you going to do with an education? You're going to get married; you're going to have children; you're going to be a housewife all your life." So it was something that was denied to me and I always wanted to further my education. . . . I saved. And when I saved during my younger days, I saved enough to go. Why I waited so long, people ask me that; I ask myself. Because I had aged and ailing parents, so even though I could've come years ago, my parents were sick and ill, and it's Asian tradition, you have to take care of them. So when my parents passed away, I told myself, "Now is the time I can do something for myself." So I decided. And in Singapore . . . people ask me, "Why didn't you further your education in Singapore?" But in Singapore, women our age don't go to school. . . . [I]n America, there's no age. You can go to school however old you are, as long as you want to pursue an education. If I go to Singapore . . . I will be like a laughing stock. . . . When I first decided to come [to the United States], I had to overcome lots of obstacles. People said . . . it was the craziest thing they'd ever heard. "Why? You have a steady job; you have a job that's going to last you the rest of your life. Why do you want to go to school? You must be out of your mind." But I told myself. . . . "I know I have a job; it's going to be steady. I've always been very good in my job." But I told myself if I live to be eighty or seventy, and I ask myself, "What if I'd done it? . . . And if I could not do it, at least I tried. But if I don't even venture, there will always be a question in my mind. . . . So that was the reason why I chose to come here. . . . I took two degrees. I took the ATA [Associate in Technical Arts) degree. I graduated with an ATA degree, and now my transfer degree, AA [Associate in Arts] degree. And I'm majoring [in] international studies. And my goal is to work with a nonprofit organization. I want to work with less fortunate; I want to work with immigrants, refugees, people who come here without the language skills. . . . When I first came, people are laughing [at me]: "Oh you're not going to make it." And because of that I began to doubt myself. . . . I come to school in my forties and it's hard. It's a new culture; it's a new language; everything is so difficult. But after my first quarter I realized that not only could I do it, but I could excel . . . I am doing very, very well. My GPA right now is 3.94, so I'm doing very well. And I've got so many awards. I am, like, *Who's Who Among Students in America* three times, and I've got so many awards for excelling and also for my participation in the college. I am in every pie. I volunteer for everything. I am involved in all activities, everything. I go to middle school, high school, elementary school and share culture and I'm involved. . . . I am so happy that I [have] done this, that I really ventured to take the first step. And I was commencement speaker. . . . Being a commencement speaker really made me a celebrity overnight. People whom I've never seen, who don't know me, came to me and said, "You don't know me but I know you." I was like, "Where did I meet you?" "Oh, I was at your speech and . . . you were talking about me; you were talking to me. I could identify," she said. These people told me that "you were such inspiration that I now know that I can do it, too." Because in my speech I said, "Don't let age or gender be a barrier or obstacle when pursuing an education. You can do what you want to do and you can reach for the stars. . . . You can do it." Because I didn't believe I could do it and I said, "Look at me now. I not only do it, I excel in it." (Lucy, student, Edmonds Community College)

Lucy's characteristics, from the trait perspective—immigrant, English as a second language, out of school for a long period, middle-aged, cultural background that discriminates against women, and the like—suggest that she is an "at risk" student for both poor performance and lack of program completion. From the behavioral perspective, she is a disadvantaged student whose motivations and efforts have led her to perform at a high level. As a result of her college experiences, her initial goal of simply participating in further education beyond high school has transformed into a more specific goal of working within a nonprofit organization after attaining advanced university degrees: "I want to work with less fortunate; I want to work with immigrants, refugees, people who come here without the language skills."

Pedro who is a student in the meat-cutting program at GateWay Community College in Phoenix, Arizona, has a different experience, beginning with his walk from Central Mexico to his work at a five-star restaurant as a chef and then to his community college program at a skills center in Phoenix. His story is more typical of undocumented immigrants who leave home for economic reasons, as contrasted with F1 visa holders such as Lucy. Sandra Dozier's study describes the pressures of full-time work and part-time school attendance, out-of-state tuition rates, and ineligibility for many aid programs that often come with undocumented status. Community colleges provide an important access point to continuing education for immigrants with few other opportunities.[5]

I'm from Mexico; I have three kids. I attended college in Mexico [north of Mexico City] for business administration and I finished eight semesters . . . I was young . . . I just got excited to come here to America and I came here [by foot]. I started working in a winery . . . north of San Francisco, California, Sonoma Valley. I started working on field grapes and eventually I find another job in a restaurant as a dishwasher. . . . Well, I started doing dishes and they gave me the opportunity to start as a salad guy. . . . I met a cook who was working in another restaurant that was a little more upscale for seafood and steaks. . . . I was there for about . . . three years. I got to be a head cook. . . . I learned a lot from him. . . . One day I went [to a resort hotel] with a friend of mine; he was filling up an application and I got an application there and they called me. That was about twelve years ago. . . . I got hired. . . . It wasn't about money anymore for me; it was about knowing more. It was about the excitement of working in a five-star hotel. You could see all these chefs wearing all their nice white cooking shirts and . . . it was one of the most prestigious places there in the late eighties or in the middle nineties. . . . We worked fifty-five to sixty-five hours a week. I found the time to go take some computer classes at the . . . community college in Santa Rosa. . . . I went there and I took some Windows 95 [course] and I finished. I took that for credits, to accumulate some units, so I finished that. I took keyboarding for another semester. I took PowerPoint and I took Excel 5.0. Always trying to get everything to be, to be better.

For Pedro, experience both in a new country and in the world of work constituted part of his personal development, which would later influence his college choices and educational attainment.

All cooking involves meat cutting. I cook a lot of meat. . . . Since I started doing dishes, I wanted, I always wanted to learn more, and I started cutting the steaks for the restaurant. A lot of seafood . . . salmon, cleaning . . . all kinds of fish, like, [I] can name a lot of them. Salmon, sole fish, mahi-mahi, tuna. . . . Cooking doesn't make money unless you're an executive chef and you have a really impressive background and work history, [then] you might be making good money. Being an executive chef . . . can take ten, fifteen, twenty years. You might not even become one, but it's not the same. I mean, if you want to make money, because you can be an executive chef of a little place, just a little place, but being an executive chef of a big-name resort will cost you a lot of years. I'm not talking just about years of work; it can cost you a lot of other things, like family relations. You don't spend time with your family.

As well, Pedro's personal experiences and the acquisition of a family led him back to his Mexican cultural heritage and the value of family.

I have a family. I actually never had a family until now [because now] I'm not cooking. . . . When I was working at . . . the kitchen for the cafeteria, they had a cook, and we hired this guy who actually was a jeweler, and he was looking for part-time, so he actually became full-time, but he was a jeweler. . . . So he went back to the hotel in California and gave notice and he came here [Phoenix]. . . . [I d]idn't know anything about jewelry, but . . . I came here and I worked. The second day that I got here I started working in jewelry. . . . I met my girlfriend a few years ago, and I know, I know because I work fifteen years in kitchen and I knew my relations wasn't going to be that strong working nights, so I didn't want to go back to kitchens because of that.

Pedro's educational choices are a combination of his life's experiences and his cultural background, and to some extent chance. His connection with GateWay Community College is a combination of his predisposition to the program and his search for information for personal growth.

One time, and not too long ago, just five months ago, I was [at] the local library. . . . I always wanted to learn more and to find a way to get a better job, which is the main thing for me to support my family and [be] secure: a better life for me. . . . I was going through the catalogs before this. I took another class at Glendale Community College [in Phoenix]. . . . I saw the GateWay Community College catalog, and second page after I opened the catalog, I see this skills center. I saw the meat cutting . . . I heard about cooking, cooking culinary schools, cooking programs, cooking seminars. . . . It got my attention . . . and I came here and I got a little more information. . . . Besides . . . the places where they can work, which is, like, a grocery store . . . not just meat cutters but most workers there get benefits that will help your family. . . . It's been really helpful; you just don't learn how to cut meat. The program breaks into a lot of other things, into meat wrapping. . . . You get a certificate as a meat-cutting apprentice. . . . I believe you can get a certificate [as a] deli worker or even meat wrapper. I'm still working at the jewelry . . . I work there and I'm attending here on my days off full-time, and on the days that I work I'm here half of the day from eight to eleven. I'm putting [in] about twenty, twenty-five hours a week. . . . I'm working fifty hours a week [at the jewelry store] . . . six days a week. I do have a family: a baby, eighteen-month baby, and an eleven-year-old girl. . . . It's hard. . . . A girlfriend: live-in girlfriend . . . she's starting to work a few days: She didn't work for a while because the baby was too small and it's not worth it working and paying a babysitter. . . . Just us and not even family here. . . . School means a lot to me. It's just really important. It's just really, really important. It can be any other area, but school is really important. At this time, if you are not prepared you are not safe. It's not just you, it's your family, and this is one of the ways, and this skills center is just changing people's lives. . . . If you know how to use it, if you really want to do something, this is one of the best places to change your life. . . . You get a management position: that would be the top, every . . . meat cutter's dream [is] to become a meat manager. . . . Well, I'm just a guy the school [gave] the opportunity to . . . and sometimes the wrong people are taking the wrong opportunity and they don't realize that, and they don't appreciate the opportunity. . . . (Pedro, student, GateWay Community College)

Pedro has changed his life several times and in dramatic fashion, initially by his journey from Mexico to California. Through his developmental process from a twenty-year-old to a man in his thirties, and through his experiences as grape picker to dishwasher to chef, he increased his value for learning and his need for personal economic security with his family. While he attributes the Skills Center at GateWay Community College to "changing people's lives," the college afforded Pedro the opportunity to change his own life. His altered life circumstances since his college experiences in Mexico shaped the choices he made, and both his tenacity and goal orientation, evident in his walk from Mexico to the Sonoma Valley in California, are no doubt connected to his over-coming of obstacles.

How students arrive at college and the paths they take within their educational journey are parts of their educational experiences, shaping their perceptions of college and its outcomes for them. In contrast to Baxter Magolda's students,[6] who begin college at age eighteen, these nontraditional students do not express themselves through their curricular passages, nor through a sequencing of mathematics courses or their maturation through reading for intellectual development. Nor are their social life or sexual experiences the markers of their development as we can note in other "coming of age" investigations.[7] Rather, these students are in the process of reshaping and redefining that life or creating a new one.

Sue recounts her background before arriving at Edmonds Community College in Washington State. Her life's experiences before education at Edmonds reflect considerable hardships and setbacks, including serious, life-threatening injury and financial debt. Her current college experiences, as both a student and as a worker on campus, are shaped by her life's experiences. Her perceptions of college are contextualized within her physical conditions, both past and present. Now, her worst fear is not doing well in her academic work; prior to college, it was dying.

> I went to college for about a year and a half after I graduated from high school and I didn't do very well . . . pretty low self-esteem, pretty different person than I am now, and [I] ended up getting married. I have two daughters, almost twenty-five and twenty-two, the oldest one is married. . . . I supported my daughters and myself for many years. And I was in an auto accident about six years ago that prevented me from doing my business full-time, so I tried other things. . . . Been training locally, regionally, and nationally for that company—and then the accident. Well, it was just a fender bender, [but] after two years left me with all the symptoms of MS [multiple sclerosis] and I could barely walk or talk. The medical doctors gave up on me, told me to get my final papers in order because I was going downhill so fast and they couldn't find out what was going on. So I had to make a decision: If I was going to live, [and] I needed to since my youngest daughter was still at home, then I needed to have quality of life. . . . I was pretty positive anyway, but there's a big difference when you deliberately get up every day and look for ways and things to be grateful for and thankful for. And by doing that, within two months, I went to a naturopath and just started acupuncture and the holistic route . . . So I was working. I was going to become a salesperson for a bathroom manufacturing, bathroom partitions [business], for like schools . . . and so they brought me in to just learn about the business as a receptionist, and then I became a self-taught project manager. . . . [I] left that job to go to work for Better Homes and Hearths, which is a wood-burning place. . . . I became a salesperson for them. . . . And then I got into an accident. I tripped on a safety net and went headfirst into it. . . . That was April of last year. Things became so unbearable at work in October, where this guy was basically threatening my life and they weren't doing what they needed to be doing. So I ended up quitting, leaving there. . . . I had to sell my home after the first accident, after three years, trying to survive. And then I was working the two jobs to try and pay off all the medical bills, and so the last accident made it impossible, and I ended up having to file bankruptcy anyway. So I'm bankrupt, I figure I don't have anything else to lose. If I'm going to do it, now's the time because I'm already broke. . . . I'm in the worker retraining program.

Sue has become a student not in spite of her physical and cognitive difficulties and not by eluding her past, but by engaging and persisting in her institution and its educational programs; as she concisely says, "by showing up."

> I learn so much every day, more than just what's in the book. I mean, my job over at services for students with disabilities, it's so humbling and I

have ADD [attention deficit disorder], which I didn't know about until about eight years ago, which probably explains why I never did really well in school before, but I'm on the honor society or whatever they call it here. I have a 3.79 grade point average, which is very exciting, because that's just totally different than what I was before.

Student experiences are also those gained from their interactions, negotiations, and learning while at college. Karen at Bakersfield College in California expresses the negotiated education she has managed to achieve. She arrived at the college with a referral from Cal Works, the state welfare program.

I'm a reentry student, I'm forty-three. I'm a single mom of six. First-generation college student. Growing up, I was never encouraged to go to college. Ended up in an abusive marriage; two years ago got out, and knew the only way we could stand on our own two feet was for me to come back to school—and had to fight to get here. Without [the Cal Works coordinator on campus], I would not have succeeded. I was a blubbering mess when I got here. I had no clue where to start, what to do or anything. And she's kind of helped guide me through the process until I was on my feet enough to do what I needed to do and be successful. [Cal Works] referred me to her because she's here on campus and she knows the ins and the outs and the steps that we need to do to get enrolled, our assessments and all of that. Because if I had, if they had just told me, "OK, you can go to school," I would not have known what the process was. And I was not in any shape to . . . because if you are coming back to college after being away, you don't always know the process or where to start or where to begin or who to ask. And when you come out of bad situations like most of the Cal Works participants, your esteem is so low anyway, that you need that extra guidance. Or at least I did. . . . I've had my ups and my downs because I still have to deal with a lot of the issues that we dealt with in our circumstances. I've had two major surgeries four days apart that I scheduled at the end of summer school so that I could make my fall semester, which I probably shouldn't have done, but I pushed it . . . I did summer school, had two major surgeries, and was back here on crutches in the fall so that I wouldn't miss a semester so that I could graduate on time . . . May 21—counting down the days. But I knew that if I missed a semester they [Cal Works] wouldn't allow me to come back and there was no way. The surgeries were a have to thing. Didn't want them, but it was my health or nothing, and I wasn't giving up school. It's an AA [Associate in Arts]. To me, there wasn't that much difference between the AA and the AS [Associate in Science], and the county only gives you a certain amount of time to get in and get out, or they cut your support. . . . I had to fight to get what I got. They were going to cut me off. I ended up starting mid-semester, which meant I got no semester-length classes my first semester because it was either "start now or you're not going." So I started the end of February, which was already halfway through. So I took whatever open-entry classes there were available just to get in. Like, "OK, you're giving me the shot, I'm taking it," and then they wanted to cut me off and weren't going to allow me [readmission in the] fall because I had the surgeries and I was off, and it's like "uh-uh" . . . I fought them. . . . And then they weren't going to give me spring and I fought them again, and I got there through May. . . . Well, sometimes it works and sometimes it doesn't, and not everybody's willing to argue to do it and I just, I had a wonderful advocate in [the coordinator here at Bakersfield]. Without her, I would've lost a lot of income. I would've still stayed in the program. I would not have dropped. But it would've made life extremely difficult. (Karen, student, Bakersfield College)

Karen's struggles were not so much that she was a returning student with children, but that she was economically and educationally dependent upon the state and subject to the decisions of local bureaucrats, a condition that is amply explored in the work of Kathleen Shaw, Sara Goldrick-Rab, Christopher Mazzeo, and Jerry Jacobs.[8] Karen's experiences are described as a battle: "I had to fight to get what I got. They were going to cut me off." This was a battle against the Cal Works program that took the position that she was not entitled to education because of the extended time component. Karen portrays her struggle as a metaphor where she challenges the monster in her quest to attain her associate's degree. She may not have slain the monster, but she defeats it and is able to press forward toward her goal.

Not all disadvantaged students triumph or persist or have positive educational experiences. Some have motives for educational participation and some have lives that are not aligned with the purposes of education. Greg, a nursing instructor at Wake Technical Community College, apprises us of some of his certificate program students.

There are a few that are here strictly because they want to keep getting welfare to fulfill that requirement that "I tried to get a job, get work." My last class I had three of them like that.

As well, there are those students whose life's experiences have not prepared them for college and whose personal circumstances influence not only their college attendance but also their college performance and attainment. These are students typically found in Greg's classes.

Most of them . . . this is that first wake-up call. I have two kids [students] now [who say] "I have suspects for fathers." "My parents have had it with me; I am twenty-plus years old." "I have to do something or I am going down the crapper." I see a lot of concern: [They are] nervous, scared, because I think for a number of them it is the first time they are going to actually take responsibility for their actions. (Greg, Nursing Instructor, Continuing Education, Wake Technical Community College)

Aside from their contact with an instructor such as Greg, these students have no other association or interactions with the institution.

At Truman College in Chicago, Ed an instructor of GED, characterizes the majority of his students and ascribes to them an identity as a person with disadvantages facing adversity.

[T]he people who enter GED by definition are people who have never finished high school. Overwhelmingly, they are people born and raised in Chicago. I have some immigrants [who], for various reasons, either didn't graduate from high school or their records are destroyed or they can't get them, because they are refugees or whatever. Typical age has become younger and younger. It use to be, back in the eighties, when I was teaching this class, my average age would to be thirty-five or forty. Now, my average age is twenty-five or twenty-eight. People drop out of high school or they never finish; they drop out for a million different reasons. Usually it has to do with kids that are poor, their friends are into drugs, they want to go out and play, there [are] alcohol or drug problems in the family, they have to go to work, they are pregnant. You know, all of those kind of things, plus a million different other stories.

Dorothy, a GED student at Truman, is also a deaf person. Dorothy is unquestionably a person with disadvantages who has a life of adversity. Her educational goals are framed by several factors—her deafness, her goal of learning to read English, her need to help her grandchild who is also deaf. It seems unlikely that Dorothy is having positive experiences or that college is rescuing her or preparing her for a new and promising life. She speaks through her interpreter.

I was born in Mississippi. I don't know what part of it . . . I don't know how old I was exactly when I moved to Chicago: maybe nine years old, when I moved to Chicago. I was married twice, have five kids. I use to work at a post office for nineteen years and then I retired in 1991. And I just became a grandmother. [I have] probably eight or nine grandkids. That is enough. . . . I go to school because I want to learn English: how to pronounce the words, and I want to teach . . . deaf people that don't know sign language that well . . . A lot of the times when I read a newspaper, like the captioning, it is a lot of words and I don't understand what it means. . . . And that really bothers me, and I wanted to know what was going on. I wanted to be able to teach to my grandkids. One of my grandkids is deaf, too. . . . I am only focusing on what I am doing. I am not trying to think about anything else but that. I am not thinking about what I will do after that. . . . Right now, I am still frustrated. That is what I am focusing on now. (Dorothy, student, Truman College)

A dean at Truman College describes a specific population of African American students who present fundamental personal problems that the college is unlikely to solve. He uses the word "oppressive" to express how he and others who work in the institution perceive this population. While the condition of this population likely does oppress them, their responses to their conditions are viewed as actions foisted upon the college, as if they are oppressing the college and its members.

> Our African Americans are . . . oppressive, and they will come and tell you they need money for the bus ride to come to class and [they need you to] help them, send them to agencies, call their caseworker, and see if we can get them more tokens. . . . We have contacted other agencies to get them more tokens. We have the ones that come out and say, "I am dyslexic." "I can't read." "I'm illiterate." These are areas you need to work on: Send them to tutors; send them to other agencies that can handle illiteracy on a one-to-one basis. We have a tutoring department and we have, I don't know how many, but ten or twelve tutors.

The social ills and conditions that surround community college students, such as those described at Truman College, are brought to the college with these students, as part of students' lifeworld. As in the example above, these orientations and cultural understandings pervade the college, and they influence interactions to the extent that faculty and administrators view these interactions as a form of oppression. That is, in this system the ostensibly oppressed population—African American students—in turn oppress the elites—administrators and faculty. This observation is in conflict with early works on community colleges where the institution's elite were viewed as the oppressors.[9]

In some contrast to the claims of institutions and the promotional literature of practitioners, community colleges are unlikely to rescue this country's disadvantaged or remake this social structure so that there is a lessening gap between the "haves" and "have-nots." Norton Grubb sees the community college as the only beacon of hope, not only for disconnected youth, but for pathways to vocations for low socioeconomic students, more so than any other institution or organization.[10] Arthur Cohen and Florence Brawer eschew the socially transforming role of community colleges; instead, they assert that community colleges do hold out the potential for individual social mobility and enable students to become effective and responsible members of society.[11] Yet the sheer magnitude of social and economic conditions for disadvantaged groups—such as low or no wages, health problems and no insurance, unhealthy and dangerous neighborhoods, dysfunctional or no families or too much of a family, and of course lack of education—is far too daunting for community colleges to overcome.

Individual circumstances and personal conditions are not simply remedied by college, as illuminated by Tereaza, a GED student, whose college experiences are shaped and structured by her family.

> My first goal is to get GED. Then after that, I don't know, maybe I will get some degree. It takes me a long time because I don't feel like I get a lot of

time to study at home . . . I used to come four days a week and I have to cut it to one day a week, because the kids will be home for the summer and I don't have anybody to watch them. I feel like I can only squeeze out one day. . . . [The college] is very close to my house. It is very convenient. I take the train and I am only, like, twenty minutes from here. There is not another place, so I chose Truman. . . . I do my studying [at the college]. I don't have any time at home. By the time I put them [the children] to sleep, I am so tired. . . . My husband will sometimes call me here and I will have to leave and cut a class to go home. . . . Classes are free, unless you have kids. Then you have to find someone to watch the babies.

Indeed, Tereaza has little perspective on the future; she is uncertain of educational goals and how to pursue or achieve them. She is oppressed by her domestic responsibilities, and her education is subordinated to her child care concerns.

Conditions for students such as Tereaza and others are not conducive to what customarily, in the literature, constitute student achievement or attainment, usually a degree or at least a certificate. Concepts such as student engagement, academic and social integration, as well as traditional notions of student development,[12] refer to experiences that are remote from these students' experiences. Indeed, there is a significant gulf between these students' experiences and those of traditional students, even those who have identity challenges as minority students or gay or lesbian students but nonetheless are traditional college students and have less extreme conditions—economic, personal, and social—than those of the students I refer to as nontraditional from the behavioral perspective. The institution, as well as public policy, faces significant problems and hurdles in order to adjust conditions for the disadvantaged students highlighted here.

Not all nontraditional students have backgrounds or conditions that merit the label "extreme" or highly disadvantaged. Indeed, younger students who are not immigrants but are recently out of high school and participating in college degree programs, while similar in their traits to traditional students, are disadvantaged, too, and seemingly struggle to persist. James is twenty years old, a former high school and present-day college football player, and he is undertaking a liberal arts academic program. His struggles are typical of those students whose academic background from high school is weak. His language skills in particular are weak, as evident in his comments.

I took the assessment test and I found out that my writing skills weren't that good, so I took a, kind of a reading class . . . to learn my reading better, because I don't really enjoy reading. So I'm very lazy about that. So it taught me how to read and take notes better, and then I took my English 60 and I learned how to write. Realized how many mistakes I made. So I took English and I took history. That was interesting. My teacher . . . taught me a lot, but it wasn't really hard. So it's more like you pick what you want to learn and write, like, a four-page essay about it. So you're learning at the same time. I really like Psychology. I like the teacher. He taught me a lot. He's also my counselor. So that worked out good. I also took speech. I like speech. That teacher made me. I want to become, like, kind of a speech teacher, too. It was real interesting in Communications because I could talk a lot, and so it's not that hard. Right now I'm taking Biology and Math; this is probably my hardest semester. Just to get my Gen Ed [general education] out of the way. Math, I'm really not that good at it. It's really hard. Get a lot of homework. . . . My girlfriend helps me. . . . My studies is better. Like when I want to study. Like when I want to learn my subject and stuff like that. I learned; I learned a lot about life that I didn't know: like a plan . . . know what you're going to do, not just run and do anything. (James, student, Bakersfield College)

James obtained a fee waiver from the institution and received a small scholarship to attend the college. He plans to transfer to a private university in the same state following his second year of college. His career plan is in fact to become a special education teacher as well as an activities director. But institutional assistance with college attendance is not solely what James needs in order to persist in college and, particularly, to move educationally in a direction that will approximate his career aspirations.

Perspectives on students' experiences of college indicate far more than the trait perspective: not only how college affects students, but also the likely educational performance and attainment of students. Traditionally, the behavioral framework has been used in ethnographic research, such as Moffat's *Coming of Age in New Jersey: College and American Culture*, and more recently, Nathan's *My Freshman Year: What a Professor Learned by Becoming a Student*.[13] But these accounts address traditional students, those far different in their backgrounds and life's experiences than the students I discuss.[14] As well, the issue in these ethnographic works and others, with the exception of Bowl's *Nontraditional Entrants to Higher Education* from the United Kingdom, is not educational performance or attainment, unless one makes an exception, too, for Holland and Eisenhart's *Educated in Romance: Women, Achievement, and College Culture*, which although focusing upon traditional students, does show the limits of educational performance and attainment among students who are members of sororities. For students who are not traditional students—because they are adults with considerable life experiences or because they come from disadvantaged backgrounds—their experiences of college include their struggles, both before and outside of college, as well as within college. These experiences are reflections not only of both student and institutional characteristics, but also student identities and institutional behaviors as colleges act to meet their own goals and strategies. These are the subjects of the next two chapters.

Student Performance: Self-Assessments

In some contrast to the views of scholars, the assertions of external observers and policymakers that address the inadequate performance or the formal institutional claims about learning achievements of community college students, the reflections and judgments of students themselves on their college experiences suggest different understandings of educational performance. Students' reflections indicate what they are learning, what results from their college experiences, and the ways in which they fit their college experiences into their "lifeworld."[15]

Whereas there is considerable debate on student performance in community colleges, that debate clearly uses institutional, rational, and bureaucratic frameworks and is largely the product of scholarship on university students.[16] Transfer rates, degree completion, and dropout patterns serve as proxy measures for both student attainment and institutional performance. These measures do not necessarily take into account such matters as student "readiness" to transfer in transfer rates; the actual behaviors consistent with degree pursuit and achievement (such as taking a program and the appropriate courses for the degree), and the learning needs and intentions of students in attending college that might redefine the meaning of "dropout."

Latrenda at Community College of Denver (CCD) serves as a useful example of the complexity not only of adult student life, but also of program and course-taking patterns.

> My name is Latrenda. . . . I am a student at . . . Community College of Denver. I will be graduating next week . . . with my degree in Applied Computer Science. . . . And I decided to change my major at the last minute, change it to Nursing. So I'm starting a whole different degree in nursing. . . . I've been a student here for about six years off and on; that's how long it's taken me to get that first degree. As I said, I take some time off raising three kids, [who] are now teenagers full-blown, eighteen, sixteen, and fifteen: I know, it's tough. But people often ask me, how do I do it? . . . Working, going to school and raising three teenagers; it is hard, it's not an easy task. . . . I made it to the twelfth grade and didn't graduate. My mom got pregnant, had another baby, and so I just stayed home a lot watching him; so I didn't graduate high school. But I did go back and get my GED. . . . I started with the basics for my GED classes, and then I started my computer classes, which were, I can't remember—that was so long ago, I can't remember. I do, I can't remember the exact courses I was taking at the time, I just remember I was pulling nineteen credit hours my very first semester at CCD. That very first year, well, that first semester I pulled nineteen credits, I was a full-time student for about two and a half years, the first two and a half years, pulling anywhere from twelve to nineteen credit hours. . . . This semester's been really tough for me because I changed over and switched from one degree, going from computers to the human body. A lot of people say it's basically the same; it's not. . . . I take, let me see, I have Psychology 235, I have Nutrition 100, and I have Biology 109, which are pretty tough. Those are pretty tough classes, Biology especially. In fact, this was the first semester since I've been a CCD student having to get a tutor for Biology, which . . . has really, really helped me.

She's also a student. . . . I pulled my grade up from an F because I was struggling so bad in the beginning, to, I'm about a B now. . . . I am at ninety-four credit hours right now with CCD, but that was for my last degree. . . . I'm starting a whole different degree now.

Latrenda moved in her academic pursuit from a GED to an associate's degree; yet instead of moving on to job placement or transfer to a university, she will remain at Community College of Denver for another associate's degree. Her reasoning for an alteration to her educational plans is based upon the labor market and the disconnection between her training and the potential for a career in the computer field. Her course-taking pattern falls outside what is expected of college students: Finish a program and move on either to work or to a higher level, such as a university or graduate school.

I have been working in the computer field for the last couple of years, or computer world, or however you want to put it, and I keep getting laid off. At the time I started this degree, they told us, "Oh . . . you'll always have a job in computers." . . . Well I'm finding that it's just not so. I keep getting laid off. Like I said, I've been laid off twice. Right now I'm not working, I work here in the first-generation student success office as an assistant, but that ends next week when the semester ends. It's just temporary. Well, I just decided to, I wanted to go into a field that I know will always be around and I know that I will always have a job in that field. I found that during the time I started this degree, they were also offering the same exact classes that we were taking for a degree, they started offering them in high schools. Well, guess what? The high school students came straight out of high school and took up all the jobs with just certifications, while we're still in school getting a degree, pursuing a degree. So by the time we finished our degree, guess what? There [are] no jobs. They're already taken. (Latrenda, student, Community College of Denver)

Her program decision—to move into nursing—was not accidental but rather the result of prior experiences as a nurse's aid. During this prolonged educational journey she maintained a 3.0 GPA at CCD. Would she have achieved even higher grades had she greater financial assistance and fewer needs to work? The reasonable answer is "yes." Furthermore, her experiences as a working adult— working at Best Buy in the services department with computers and previously as a nurse's aid—shaped her educational choices, contributing to her program intentions. Yet, in spite of her abilities and achievements, Latrenda lacked the social capital and personal confidence to ensure that she was accorded the grades she deserved or could have attained if given the opportunity.

I was working full-time and I was going to school in the evenings from five to eight-thirty at night, and I encountered some unfortunate things that happened in my family. I had three deaths, like, back to back, and [there] were self-paced classes and I fell a little behind. . . . I was just stressed and I went to the instructor and I explained what was going on to him in my personal life: "What can we do? I know I've fallen behind, but what can we do? I don't want to end up with a Fs on my transcripts." . . . Well, he explained . . . that would be fine, and this and that, and he understood and everything and I told him, "OK, I'm just going to have to chuck this semester up. I'll deal with financial aid and everything else later, next semester," and we talked about him giving me incompletes. . . . When I got my grades, he had given me two Fs and my feelings were hurt because I just, even though I went to him personally, because that's what they ask you to do, we sat down and talked about it, and he told me he would give me incompletes and still ended up giving me Fs, and that just really, I didn't think that was too kosher, especially with us talking about it. . . . I had to retake the classes, of course. . . . I left him messages and he never returned my call. (Latrenda, student, Community College of Denver)

For non-traditional students, both their experience of college and their performance in courses and programs are complicated. The factors that have a bearing on their program decisions are various, including prior work and educational experience, their socioeconomic background, and their ethnic origins, as well as their family responsibilities. The judgment of these students' educational performance and attainment, using such measures as GPA, transfer, time-to-degree, and persistence generally, is not appropriate for understanding how students experience college and how college might affect students.

Notes

1. Marion Bowl, *Nontraditional Entrants to Higher Education* (Stoke on Trent, UK: Trentham Books, 2003).
2. John Rawls, *Political Liberalism* (New York: Columbia University Press, 1993); John Rawls, *A Theory of Justice* (Cambridge, MA: Belknap Press of Harvard University Press, 1999).
3. Marcia Baxter Magolda, *Making Their Own Way: Narrative for Transforming Higher Education to Promote Self-Development* (Sterling, VA: Stylus, 2001).
4. Sandra Bygrave Dozier, "Undocumented and Documented International Students: A Comparative Study of Their Academic Performance," *Community College Review* 29, no. 2 (2001): 43–53.
5. Ibid.
6. Baxter Magolda, *Making Their Own Way*.
7. Dorothy C. Holland and Margaret A. Eisenhart, *Educated in Romance: Women, Achievement, and College Culture* (Chicago: University of Chicago Press, 1990); Michael Moffat, *Coming of Age in New Jersey: College and American Culture* (New Brunswick, NJ: Rutgers University Press, 1989).
8. Kathleen Shaw et al., "Putting Poor People to Work: How the Work-First Ideology Eroded College Access for the Poor" (unpublished manuscript, 2005).

9. Robert Rhoads and James Valadez, *Democracy, Multiculturalism, and the Community College* (New York: Garland, 1996); Lois Weis, *Between Two Worlds: Black Students in an Urban Community College* (Boston: Routledge and Kegan Paul, 1985).

10. Norton Grubb, "Using Community Colleges to Reconnect Disconnected Youth" (Menlo Park, CA: William and Flora Hewlett Foundation, 2003); W. Norton Grubb and Marvin Lazerson, *The Education Gospel* (Cambridge, MA: Harvard University Press, 2004).

11. Arthur Cohen and Florence Brawer, *The American Community College* (San Francisco: Jossey-Bass, 2003).

12. Ernest T. Pascarella and Patrick Terenzini, *How College Affects Students: A Third Decade of Research* (San Francisco: Jossey-Bass, 2005).

13. Rebekah Nathan, *My Freshman Year: What a Professor Learned by Becoming a Student* (Ithaca, NY: Cornell University Press, 2005).

14. The exception is that Nathan is herself a nontraditional student who, as a university professor, goes undercover as a student to provide an ethnography of student experience.

15. Dorothy C. Holland et al., *Identity and Agency in Cultural Worlds* (Cambridge, MA: Harvard University Press, 1998).

16. Thomas Bailey, Davis Jenkins, and Timothy Leinbach, *Is Student Success Labeled Institutional Failure? Student Goals and Graduation Rates in the Accountability Debate at Community Colleges* (New York: Community College Research Center, Teachers College, Columbia University, 2005); Thomas R. Bailey and Mariana Alfonso, *Paths to Persistence: An Analysis of Research on Program Effectiveness at Community Colleges* (New York: Teachers College, Columbia University, 2005), 1–70; Linda Hagedorn, *Traveling Successfully on the Community College Pathway: The Research and Findings of the Transfer and Retention of Urban Community College Students Project* (Los Angeles: University of Southern California, Rossier School of Education, and University of Florida, 2006); Pascarella and Terenzini, *How College Affects Students*.

Chapter 5

Student Characteristics and Institutional Contexts: The Nexus of Student Experience and Attainment

The previous chapters have laid the foundation for understanding how new nontraditional students experience community college, what goals and motivations direct their activity, and how the colleges respond to them with their strategies and, ultimately, actions. This chapter will examine the question, "Are students receiving just treatment?" in different college contexts and across differing student populations within the nontraditional categories. I take three community colleges and compare student behaviors and institutional actions with respect to students, and then identify both outcomes and potential outcomes for students.[1] There are both clear and dramatic differences in this comparison. These differences are in large part the consequence of two main factors: student characteristics and institutional behaviors. Student characteristics, however, are not simply what students bring in to the institutions, such as their ethnic and racial labels or their academic limitations, which might include their nonnative English language skills. Student characteristics also include what Kate Shaw has noted as the complex identity of community college students, an identity that is largely self-constructed.[2] One component can be referred to as "actual" identity; the other, "designated" identity.[3] Actual identities are those we recognize as the actual condition of individuals, whereas designated identities are those conditions or states that are expected—either in the present or in the future. Institutional behaviors include those programs and institutional actors, such as faculty, staff, and administrators, that form part of the students' experiences and development while they are at the college. Together, student identity and institutional behaviors combine to influence how students progress in their education and provide both outcomes and potential for specific outcomes for these students. These outcomes may include dropping out or "stopping out" (taking time out and then returning) of college or transferring to a university or job placement.[4]

At each college, I examine three conditions: 1) the institutional experiences of students during their educational programs, 2) student goals and related actions, and 3) the role of the community college in both education and employment. I discuss Wake Technical Community College and Harry Truman College first; then I move on to discuss Bakersfield College.[5] Through this structure and comparison I intend to show how the combination of student characteristics and institutional behaviors work to affect both student experiences and likely student outcomes. I also show how different behaviors toward students—including institutional climate and actions—affect student outcomes.

The three colleges—Wake Technical Community College in North Carolina, Harry Truman College in Illinois, and Bakersfield College in California—house similar programs and similar students. At two of the community colleges, Wake Technical Community College and Harry Truman College, a similar population within the Adult Basic Education and English as a second language programs was the subject of investigation. These students are clearly highly nontraditional students: The majority of them had already failed to complete previous educational programs and had no other opportunity for remediation or skills development; others were immigrants and were not native speakers of English. At the third college, Bakersfield College, while the student population examined certainly falls under the nontraditional category, these students demonstrate accomplishments in educational programs that suggest they have gained higher educational attainment than what the literature tells us about highly nontraditional students. The educational attainment of the Bakersfield College students is not only a consequence of student characteristics, such as their motivation and their behaviors, but also institutional effects such as policies and practices.

The differences in these two populations are evident in the voices and experiences of the students at each institution. Using narrative analysis, I rely upon the stories that the students voice, as well as the stories that faculty and administrators relate about these students. These stories reveal both actual and designated identities of students, pointing out clear distinctions between the two populations. These differences show how easily one group—which I refer to as "beyond the margins" students—is largely peripheral to organizational resources and attention, and how by implication they are a neglected population within the larger society. Many of them faced daily challenges such as improving their skills in English, maintaining a job and paying bills, or caring for a family as a single parent. Although almost all students at the three colleges were required or recommended to take Basic Education classes (e.g., remedial math, reading, or writing, or basic ESL), students at Bakersfield College were integrated as much as possible into the structure and flow of the campus, and those at Truman and Wake were treated as a separate unit with little contact with and connection to the main student body of the institution. Therefore, although the students at the three colleges began their college enrollment with actual identities that would classify them as "at risk" students, one group of students—those at Truman and Wake—were viewed by institutional officials as peripheral to the academic transfer and occupational program portion of their campuses.

College Descriptions[6]

Wake Technical Community College in Raleigh, North Carolina, enrolls approximately 15,000 students in "curriculum" (or for-credit) courses. The number of students enrolled in noncredit basic and remedial education and English as a second language classes is not tracked, but it is estimated to be approximately 20,000 students. Wake Technical Community College has one main campus, four satellite campuses, and forty-three community sites. The sites are dispersed throughout the community, which includes urban sections of the capital city and more rural areas within the county. The county has an unemployment rate that is slightly lower than the national average and a relatively low poverty rate, although the students in the basic skills programs are more likely to experience barriers and difficulties than the average resident. The basic skills division houses most of the highly nontraditional and "beyond the margins" students in one of the satellite campuses. This site is an old building within the city and is neither distinguished nor attractive. The basic skills division at Wake Technical Community College is composed predominately of Latino and African American students, although Wake County's demographics include a population that is categorized as over 70 percent white.

Harry Truman College in Chicago enrolls approximately 33,000 students in both credit and noncredit courses, with 62 percent of those students in basic skills or English as a second language (ESL) courses. Harry Truman College is one of seven colleges of the Chicago Community College System. Harry Truman College has one main site for all students. The institution is in the center of a depressed area of the city, and it stands out as a pristine element of the community. According to students and institutional staff, Harry Truman represents hope to the local area. The majority of highly nontraditional and "beyond the margins" students at this institution are enrolled in Adult High School, GED, or ESL classes. College enrollments are predominately in noncredit courses, with 22,229 listed as adult education, and the majority of these are ESL, which is not unreasonable given that Chicago's population includes approximately 21 percent first-generation immigrants. The main racial/ethnic categorization of students at Harry Truman is Latino. The Chicago population has more racial and ethnic diversity than Wake County, including 47 percent whites, 20 percent Latinos, 26 percent African Americans, and 5 percent Asians. As a large city, Chicago also has unemployment and poverty rates that are higher than those in Wake County. Historically, 30 percent of the credit students have previously enrolled in noncredit courses. In spite of the success of the institution's noncredit programs, Harry Truman College is not viewed by organizational members as advancing either in resources or its potential. No doubt this pessimism is a consequence of present conditions. According to administrators, in 2004, the institution struggled financially, was in the midst of major staff cuts, and was burdened by the state and city to meet accountability requirements. Faculty and staff indicated they were extending themselves to meet the needs of the students. In spite of the uncertainty the students and staff stated about the future of their programs, they remain relatively optimistic that the programs will have a positive impact on the future of these students. In the face of these institutional concerns and the conditions of the surrounding community, these students have developed a strong academic and supportive community with one another.

Bakersfield College enrolls 16,000 students in curriculum, or credit, classes and approximately 1,609 students in noncredit classes. The relatively low number of noncredit students is a consequence of the college's designating almost all of its courses as credit bearing. Bakersfield College has one main site, with two satellite campuses. Bakersfield is a midsize city in California, but the area surrounding it is rural, drawing large numbers of migrant farm laborers. The county has the highest unemployment rate of any in this study, at 8.1 percent. The percentage of persons living in poverty (approximately 20 percent) is similar to New York City or Los Angeles, in spite of the vastly different landscape. The institution is predominately Latino and white, with the Latino population steadily rising as the white population falls. The institution has been a fixture of the community for over seventy years and has developed strong community connections and services in a number of areas. Services include Cal Works (a program to assist those on welfare in obtaining an education and finding a job), community education, corporate and community services, environmental training, small business development, Tech Prep, and workplace skills and training. The institution also has an expressed commitment to maintain strong student services aimed at a diverse population, yet budget cuts in the early 2000s diminished actions that supported students. Almost all of the Bakersfield College students were enrolled in credit courses, even if they were taking basic or remedial education courses, and the majority were program affiliated (such as academic or occupational), heading toward a credential or transfer to a university.

Similarities across the Sites

The students at all three community colleges—Harry Truman College, Wake Technical Community College, and Bakersfield College— shared a number of similarities in their perceptions and experiences. Not only have a large proportion of these students previously failed in other educational institutions and battled with low self-esteem, they also have a number of obstacles to overcome while pursuing their educational goals. No doubt, the programs at the community colleges benefited students at all three sites.

The students stated that their achievements would not have been possible had it not been for the flexibility of the program, the direction and guidance they received about the programs, and the availability of varying types of programs, certificates, and classes. The faculty and staff were credited with having significant influence on each individual student interviewed at the three sites. Every student could mention an institutional employee who had given them assistance or encouraged them to achieve their goals.

A Wake student, Val, confirms this view of a faculty that is supportive of student learning as well as student well-being.

Teachers, they are willing to help if you are willing to learn. . . . They really do care. If you haven't been here in a while they ask you how you have been, how your family has been, and stuff like that. They are really open-minded and caring. . . . That is my experience. They are real caring. . . . They teach you like adults and not like little kids. (Val, Wake Technical Community College)

Both at Harry Truman and Bakersfield, students acknowledge the work of their teachers:

[T]he teachers . . . are always helpful and they're always there to help, so there's no way you can complain. You can't say the teachers . . . here are not helping very well: They make all that time [in] their office hours to help us. (Michael, Bakersfield College)

Yeah, yeah, the teachers are good. They are every time, so good, they talk to you like a friend and . . . sometimes, [if] we are late for classes . . . the teachers say, hey, where were you, why were you late to class? They say you gotta do this and you gotta do that and they help every time. They help; they are good teachers. (Jorge, Harry Truman College, ESL student)

The efforts of individual faculty and staff members and the learning environment of the community college make a significant impact on these students. An ESL and prenursing student at Bakersfield College, Nidia, comments on what she has gained from college: "I didn't get a good grade in it [academic course], but I learned how to find myself in that class actually. . . . I developed a lot of self-confidence." This positive impact is evident through the achievements of students at all three community colleges. Those achievements include not only program completion but also the acquisition of knowledge and self-development. Michael, an administration of justice student at Bakersfield College who is a nonnative English language speaker, praises his academic program: "The program itself, it's just educating me more and giving me more knowledge and of . . . things I didn't even know yet, and it's just a real good program." James, a liberal arts student at Bakersfield College, describes his learning achievements at college: "I learned a lot about life that I didn't know, like [having] a plan . . . know what you're going to do; not just run and do anything." A reentry student at Bakersfield College, Karen, states: "I'm learning so much. . . . Just different ways of solving problems, dealing with people and their issues." Numerous students had failed in previous educational pursuits before enrolling at their current institution. The students had attempted education at other venues: traditional high schools, four-year colleges and universities, adult education programs, and even other community colleges. At these institutions they failed to accomplish their academic goals.

Undoubtedly, because of their previous failures, many of the students started at the community college with low self-esteem or self-doubts. The students at the three sites gained confidence through their educational achievements. One distinct difference among the three sets of college students was that the students at Bakersfield College overcame their self-doubts more quickly—they were confident, motivated, and willing to take charge of their intended actions. Several of these obstacles are overcome through the support of the institution's structure and the assistance of faculty, staff, and other external structures designed to assist these student populations, including family members and friends.

ESL students often had their own characteristics separate from the other students interviewed. The difference between the ESL students and the other students interviewed rested on educational attainment. Many of the ESL students had already completed some postsecondary education or had held a high-salary or high-wage position in their home country. Despite these differences, the ESL students shared a common struggle with all community college students: to find a place in American society. The language barrier these students experience negated much of the social and economic value of their past education. The ESL students understood that they must learn English before they could utilize their past accomplishments and become active members in American society and the workforce.

Differences across the Sites

The students at the three community colleges also exhibited a number of differences. Most of these differences can be noted between the students at Bakersfield College and the other two sites—Harry Truman College and Wake Technical Community College. The differences in the students, their experiences, and their perceptions at each site are a consequence of the variations among the sites. In order to convey these differences, the following discussion is organized around three themes: 1) the institutional experiences of students during their educational programs, 2) student goals and related actions, and 3) the role of the community college in both education and employment.

Student Experiences in Their Educational Program: Wake and Truman

Students at Wake and Truman have similar stories to tell about their past educational experiences and their confidence in their own learning abilities. Given that these students have limited experience in postsecondary education and generally less than satisfactory educational accomplishments in high school, their views of the community college's basic skills courses are positive.

Students experience their Basic Education and English as a second language programs as nourishing, supportive, and skill developing; yet they also make this judgment within the context of cost and benefits. Namish, an ESL student at Harry Truman, remarks, "I don't think I and other students can afford paying for these classes. I understand that not all classes can be for free, but now it is good for us being free." Stephanie, an Adult High School student at Harry Truman, adds, "I sure couldn't afford it. If you had to pay for it, I sure couldn't afford it, I couldn't even start." The students acknowledge the low costs of their education—including no tuition—and that the schedule is flexible enough that they can work at their own pace, and that the institution does not resemble high school.

The students also comment on the relationships they experience with each other. Olivia, an immigrant and ESL student at Wake, expresses the bonds she shares with others.

> With people from all over the world and you are here for the same thing and you don't speak the same language, but when you are here, it is like you speak the same language, because you have to help each other, because you are solitary with each other. It is very nice. You make friends with people very different from you. [H]ere we seem the same.

A Harry Truman ESL student from Poland, Katya, describes a similar experience.

> My best experience? I think it is that I can meet the people. I can meet them on the street, but here I am closer to them and they are talking sometimes, and very often . . . they are in the classes, [and] sometimes when we talk about something . . . they talk about their countries; that's interesting. Yeah, it is really . . . a great experience because I like to hear about the different countries.

Yet, in spite of scholarly and research evidence that supports community learning in postsecondary education,[7] especially among those students in programs such as Basic Skills, these programs do not emphasize group work. Kelly, a GED student at Wake, elaborates: "It is not like there is any group work or anything like that. It is mostly just you doing what you need to do." The solitary learning environment is captured by Shantia, who says, "I don't talk to anyone, it is quiet," and by Angela, who says, "In any room I might be the only one doing world history and other people might be doing math or something. It is just me: independent." This results in a sense of isolation, as Kelly remarks: "Everybody comes and goes at their own pace. Sometimes it is really hard. I have spoken to people, but I haven't developed any relationship with anyone." Truman students also have a tendency to study alone, partly because of busy work schedules and partly because of the various levels of understanding that the students have. A GED student, Madeline, states, "When you have, like, let's say twenty-eight students, and each is at different levels, it is pretty hard to maintain." Szeda, an Adult High School student, relates, "I just don't talk to [the other students]. When I come to class I just listen to the teacher and when she gives me something to do."

Although electronic technology is proclaimed as both a growing and necessary tool in education and the workplace, in Basic Education programs and courses at both Truman and Wake, there was little evidence of either training or use of basic technology. The use of electronic technology to enhance instruction is also not apparent. Students rarely use computers in their studies or learn to use computers more effectively. Most students only use computers once a week. Tia, an ESL student, remarks, "Maybe, [learning] more about, [becoming] more involved with the computer, more accustomed to it. . . . That would be very helpful. I think it would be helpful to others in my situation as well." An ESL student, Jorge, notes, "Yes, we have one day we go to computers downstairs in basement and use them, and they help, they help much."

Indeed, workplace skills and workplace orientation are noticeably absent in these programs. Faculty do not necessarily direct students to focus on vocational or employment preparation. Instead, students indicate that faculty members encourage them to pursue their own interests. Angela, in Adult High School at Wake, remarks, "It is kind of just whatever you're interested in. If you want to work they would show you how to do that; if I want to go to college, they will help you with that." The Dean of Adult Education at Truman also makes it clear that students should have a variety of choices for study.

> I remember when I first came, it was strictly life skills, and then more educated immigrants came in and we had to refocus our curriculum more academically. We are now trying to create another track—the academic track—and we are very fortunate that the college is all inclusive here, so they see us as feeders to the vocational programs, the credit programs.

In spite of a nourishing and caring environment, approximately half of the students who start the classes fail to complete the class at Truman or Wake. Rafael, an ESL student at Harry Truman, states, "When they start the classes it is maybe thirty-five people, and in between that is only ten or twenty less than when they started." Another ESL student, Joseph, concurs: "When we started school there [were] about thirty-five . . . something like that, and at the end of the class there are only like fifteen, twenty. Few people left, I don't know why." A vocational student at Wake, Wong, comments that "there were a lot of students when I started. It was full, and only half, I think maybe half or one-third of the students, made second semester."

This performance stems from a number of conditions. For students with children it is virtually impossible to work full-time and attend college or go to school while paying for child care. Students are aware of the various obstacles such as the financial costs of attending college that they and other students must overcome to achieve their goals. They realize that students do fail and have failed to complete their educational pursuits.

Each student has their own story of personal challenges and obstacles to overcome in order to pursue education. Carlos, an administrator at Truman, explains the students' obstacles, in this case punctuated by images of abuse.

> They are struggling through working twenty-four hours; they come here tired. It is a big sacrifice for these students, and those are the challenges that we wish there were ways we could help them. . . . We don't know all their issues, but we know a lot of issues. There are people that come to

class who are hungry, battered wives, with whole mess of problems—everything. I have seen everything: suicides, depression.

One Truman student, Joseph, an ESL student, corroborates this perception: "A lot of these people have to work, work at nights, and sometimes it's hard. Sometimes it is hard for me to get up early because of work." Jeff, a Wake administrator, expands on the condition of these students. "We are dealing with the poorest of the poor," he says, "with the handicapped, with learning disabled people . . . most underfunded, underskilled." Administrators understand that many students are unable to stay in school or that they are out of work for an extended time period.

Part of the problem of course completion is connected to the extreme difficulty students have obtaining help outside the classroom. Institutional support staff is limited for these students.

> I wait[ed] for the counselor before. . . . [I]t is too hard and crowded over here. You have to make appointment. You had to go a day ahead. I asked for help right now and they don't have any more. They are busy right now and they can't right now. . . . So, other people they don't tell you anything. They don't say, "You can take this or that." They don't tell that. (Rafael, ESL student, Harry Truman)

That is to say, advice is not readily available for these students at Truman, and thus students are uninformed about appropriate courses and educational pathways.

Administrators corroborate the student comments by noting that student services are indeed being cut and thus are more difficult for students to obtain. Martha, an administrator at Truman, comments, "I would like to serve them better. Well, serving them is pretty much bare bones at this point. The more people they've laid off, the more nuts and bolts, no ups and no extras, my job has become." She adds, "A lot of that support staff vanished because of budget cuts. . . . But you are talking about a population that has been badly served in the past, by the system that ultimately fails them and continues to be badly served here. It is just the resources aren't here."

Further complicating the problem of completion and achievement, students are unaware of the existence of support services or staff, and for those who are aware, they rarely ask for assistance. In response to the question, "Have you talked with anyone about that [problem] right here?" an ESL student at Truman, Tatiana, replies, "No, I haven't tried, because everybody says, first you have to do GED. Finish that and then you can go." Asked, "Have you talked with any of the advisors or counselors here?" Tatiana notes, "No, I just come here and try to study. I don't have time to ask."

Additionally, federal policy, interpreted by state officials and local bureaucrats and enacted by the colleges themselves, affects student experiences and outcomes. This condition is evident in the welfare-to-work program students who enroll in community colleges. At Wake Technical Community College, the program administrator for Basic Education comments and notes the inadequacies of the policy in practice.

> [T]he welfare reform programs and everything [that follows from these in education] will only get worse. It is "get to work quick." Not, "let's train you in a trade so that you are going to be able to support your family." But "we want to get you off the road and we want to get them out as quickly as possible." They want a quick fix, but they find out they [the clients] still cannot support their family. Then what are they working for? There is a group who has limited social services benefits, like after they get a job I think they can stay there for a year; after that year is up, they are going to have to quit work, because they can't afford to work and pay for someone to take care of their kids. They don't make enough to work and keep their kids in day care. (Program administrator, Wake Technical Community College)

Overall, student experiences suggest that students possess both little understanding of the learning process and limited self-confidence. The students in GED and Adult High School courses expressed only vague understanding of what they were learning and doing while in their classes. The students often answered open-ended questions with "yes" or "no" answers. Susan, an Adult High School student at Harry Truman, demonstrates this vagueness and lack of depth.

> We read and we do math, and then we do—what other stuff do we do? She [teacher] reads a book. She comes one by one and she will read for us. . . . They help me to know words and stuff. They help me. They come to me one-on-one and talk to me like that.

As well, students, especially those with limited education, doubted their ability to succeed as they embarked on coursework. A GED student, Joan, comments, "I want to get that feeling, I can do it, I can do it, I can do it, I can do that. Right now, I am still frustrated." Stephanie, an Adult High School student, concurs: "Sometimes I think I doubt myself too, so when teacher tell me that [I did well] . . . I'm, like, are you sure? I tell myself, are you sure?" These examples suggest that these students' designated identity is limited to basic and immediate needs and reflective of how the world outside views them—as peripheral to the social and economic life of their communities and nation.

Student Experiences in Their Educational Program: Bakersfield College

The students interviewed at Bakersfield College have some points of similarity with those at Wake and Truman; that is, similar circumstances and experiences that were present at Wake and Truman can be documented at Bakersfield. But there are significant differences. In addition to the daily challenges of work, family, and survival, many Bakersfield nontraditional students have immigrant backgrounds, and many exhibited the need for remediation before proceeding into college-level course work. Diane, a coordinator in the welfare-to-work program (Cal Works), confirms the challenges for students. "Housing is a big issue for students; transportation is a major issue," she says. "It doesn't take a rocket scientist to figure out that education and training are needed by these students. We need to spend the up-front dollars on that, as far as I'm concerned, or we will pay later." A Basic Skills learning center faculty member, Stephanie, states, "The majority of the students are low SES [socioeconomic status] and work . . . and, yes, they will likely fill the low, service sector jobs in society." Michael, an administration of justice student, comments on his student colleagues.

> Students don't think college is for them. They believe it's for smarter people or that they can't do it because they're not that smart . . . . They think they can't do it. [They] struggled in high school and they think they'll struggle more, far more . . . than they did in high school and they didn't believe that they were college material.

Thus, Bakersfield College students face adversity, self-doubt, and insufficient academic skills to ensure that they can learn and progress at college. But their experiences at Bakersfield indicate that through a combination of their tenacity and the institution's treatment of them, they are "college material" and can benefit from postsecondary education.

Students at Bakersfield College, even those who are highly nontraditional students, are integrated into the institution and into the mainstream of academic and occupational programs, even if they are taking remedial or developmental courses. Nontraditional students at Wake or Truman taking English as a second language classes, Adult High School, or GED experienced their education isolated from major components of the institution, separate from academic and occupational programs. The Bakersfield students were more fully integrated and demonstrated engagement in the institution. Grace, a student from Mexico who had acquired little English when she came to campus, expresses her initial introduction to college: "Well, when I came here I just knew how to say 'hello, how are you, nice to meet you,' you know, the formal stuff. And then . . . my friends just only speak English. So . . . with them I practice." After attending orientation and undertaking placement tests at Bakersfield, Grace had a plan for coursework.

> The first semester I took just English, reading, and health, and this semester I'm taking public speaking, reading, English [some others as well]. . . . (Grace, Bakersfield College)

Although Grace had graduated from high school in Mexico and only knew elementary English when coming to the United States, she received encouragement to continue with her ESL studies and incorporate other disciplines at the same time. Her exposure to courses such as public speaking and political science allowed her a broader set of experiences on the campus. In a short period of time, Grace is able to articulate her knowledge.

> It's [political science course] very much about the government, how we have been ruled by the law, and even though we don't know this, everything government, everything [in] our lives is ruled by the government. . . . Well, it's interesting because it's very much like Mexico. . . . Mexico has three powers; here they do, too. And it's like, oh, yeah, but, and next thing . . . you know the history, why they did that, and why they think that way.

Grace's knowledge and her newfound interest in political science leads her to discuss a possible future career of working for a consulate or an embassy. The chance to take a political science class has allowed her to formulate this idea earlier than she might have in an all-ESL curriculum.

At Bakersfield College, I observed a College Development class that was intended to provide a subpopulation of nontraditional students with college navigational skills and encourage career planning. A number of the students had observable disabilities; others spoke about the barriers they had overcome to be on campus. However, they also talked about their goals to obtain an associate's degree and perhaps transfer to a baccalaureate college. These students expressed high levels of motivation and reflected as a group on the importance of maintaining confidence and moving toward goals. Their conversation was not about settling for a vocational certificate just to find a better job, but about reaching their potential. Although these students were not entirely realistic about their prospects, they did express specific ambitions. They did not set their sights at a low level and then accept the limits of their designated identities or a socially ascribed identity. The majority of students in the class believed they were going to complete an associate's degree; at least they were headed in that direction. A number of them were going into human services, one individual into drug rehabilitation, another into health care to help older people. These students assumed that they were going to transfer to a university, the majority to California State University at Bakersfield. No doubt, the road for them would be difficult and likely frustrating. Without this particular class and without their instructor who was guiding them, these students would be at sea, unable to cope with the difficulties of college life. Almost every one of the fifteen students had a story to tell. About half of them had some form of physical impairment. One student was seated in an electrical wheelchair and probably close to quadriplegia. Another student, a woman probably in her fifties, had trouble walking. And a third student was African American with a cane, probably around fifty, too. He had some difficulty walking on one leg, which he said he had injured; he also suffered an accident and had incurred brain damage. Wearing a singlet, another student—male, in his mid-twenties, formerly from the University of Arkansas at Fort Smith who played football and had a body of tattoos—said he was at college essentially to complete what he had not completed at University of Arkansas, and that he had a variety of thoughts in his mind about what he wanted to take. He appeared to have some form of learning disorder and was not coherent in his expression. This was certainly a diverse student population; nonetheless, all the students were similar in their vulnerabilities and disabilities.

On the one hand, these students are likely to achieve some form of education or liberal education, even though they may not progress to advanced levels. On the other hand, they are students experiencing a form of dignity and some amelioration of their disadvantages. They are treated as students, not as people who are outside the pale. When I met with these students at Bakersfield, they all nodded when I mentioned that confidence was an important factor in what they were doing, and several of them said, "Yes, indeed, that's it. We're building our confidence. We've got to be positive about what we're doing."

Student Goals and Actions: Wake Technical Community College and Harry Truman College

For students at these colleges, both education and a job constitute personal goals, but education is a goal because students believe it is their only means to obtain a job. The students are uncertain of what education they need and they are given little direction on how to attain goals pertaining to education and work. The students were unable to identify what courses are necessary to complete, even when they articulate a specific educational goal. Angela, an Adult High School student at Wake Tech, states, "I want to go to college and you have to finish high school to go to college." Students state a desire to work, and many voiced an area of interest, but few had an understanding of what was needed to be able to obtain a position in their field of interest. The students articulated a career or educational plan, but they did not investigate matters further or put their plan into action. Gerry, a vocational student at Wake, notes, "I've never really asked for advice on where I should go work or anything." Most students at Harry Truman, except the highly educated ESL students, are unclear about what they want to accomplish. If they do have goals, they often are unclear about how to achieve them. One ESL student, Jorge, when asked how he plans to accomplish his goals responds, "I don't have any ideas about that." Antone, an ESL student, has a similar response: "I am not sure. I can't decide. I am telling myself, so far so good: in the future, maybe next year." The students are also slow in taking steps to pursue their goals. Those in ESL, who already had some postsecondary education, articulated goals that were tied to economic and social outcomes. Students expressed frustration when they perceived that their academic work imposed on their employment. Although these students indicated that education is valuable for job attainment and economic stability, they lacked a detailed goal attainment strategy, as well as any action plan in pursuit of those goals.

A handful of students were swayed into taking specific steps to pursue goals, if they were given a suggestion or specific instruction. Wong, a Wake vocational student, stated, "A guy encouraged me to start this program. He's lazy, but he guided me this way and so I have no complaints." When the students are given a suggestion or idea on how to accomplish a goal, they indicate that often they would accept and ultimately follow the advice. Todd, a GED student, remarks, "A counselor helped me figure it out. The first day I talked to her, I talked to her for two hours and she spent a long time [with me]. That is how I came up with the things that I am looking at studying." Those who were given explicit and forceful advice followed that advice.

Students had limited understanding of and experience in the workforce: Their knowledge of the economy and skills needed for the workplace was limited. Those students who sought vocational education often had a clearer idea of the education necessary and an understanding of the field they were pursuing, but they possessed few tangible ideas of what they wanted to accomplish once they finished school and secured a job. The vocational students also understood that U.S. jobs were becoming more global, outsourced to other countries. Stanley, a Wake vocational student, remarks: "Yeah, I could have gotten a job if I wanted to move from the States." He was pursuing education to find a service sector job, but understands that "at my age you may not get a job anywhere." Gerry, another vocational student, agrees: "I always wanted my own business and I may one day, but I think for now I just want to get a good head start. Maybe they won't sell out to Mexico or somewhere, where everything else is going."

Students in GED and Adult High School programs had limited knowledge of how American society operates with regards to work and money. These students also did not understand what credentials are needed to fill certain job roles, and for those that did, they were unaware of how to obtain those credentials. On the one hand, students did not realize all the possibilities for work and employment; on the other hand, several students articulated goals that would likely be impossible to fulfill considering their educational backgrounds and academic skills.

In coping with their present and their anticipated futures, these students demonstrated a high level of dependency upon college staff or family members or those with whom they have close relationships for direction. This dependency is based upon a lack of certainty about their own abilities, needs, and specific courses of action they must undertake so that they can attain goals. Overall, the students had difficulty articulating their thoughts and ideas. Their educational experiences were not well integrated into their thought processes and were not clearly connected to their future plans or ambitions.

Student Goals and Actions: Bakersfield College

Students at Bakersfield College understand that they must take risks and make sacrifices to achieve their goals. Maria, formerly in the Army, exemplifies this pattern.

> I thought that was a sacrifice to go four years [in the Army] and in return I can go to college, because that is what I wanted to do. I wanted to make a better life than what my parents gave to me and just move up the chain. So, I started here in the community college using my GI Bill, but at that point that wasn't enough money. So, I worked and got financial aid at the same time. (Maria, applied science and technology student)

Furthermore, there is evidence of connections between work experiences, educational goals, and curriculum choices in the population of nontraditional students at Bakersfield College. Lidia is preparing to apply to the nursing program at Bakersfield by completing a long list of science prerequisites at the college. In addition, she had to complete some English classes as a nonnative speaker. However, her current job in a hospital motivates her to pursue this arduous educational path.

> I work in a hospital, actually labor and delivery. My job there is O.B. Tech. I'm the kind of person that prepares the O.R. for the doctors and help them get what they need. . . . [I learned that] right there in the hospital. They trained me there. I went for the month of training. I went through the main O.R.; they showed me how to wash your hands, how to clothe the doctors. Then I went two weeks to the sterile technique, all the instruments and stuff like that. . . . I'm working in nursing in the medical field already. I like to take care of people. I just, I feel I have that gift, that give[s] me that gift for caring and generous and loving. That's what I think. I don't know. I always wanted to be a nurse. (Lidia, Bakersfield College)

In comparison to some of the ESL students at the other two institutions, Lidia appears to be more focused and aware of the steps needed to reach her goal.

The Bakersfield College students understand that obtaining a certificate or an associate's degree will not solve their problems, but nonetheless is an important step in their preferred direction. They realize that there are more obstacles to overcome, as expressed by Karen, a reentry student and mother of six.

> We're never too old to stop learning. . . . There is a lot out there to learn; there's a lot of different possibilities that we don't always consider. I know coming out of high school you don't always understand all the different possibilities as far as jobs and where you can go and what you can do with it; it opens up a lot of worlds for us.

Diane, coordinator of the Cal Works welfare-to-work program for students, confirms the desire for students to develop this mind-set.

> One of my missions is to take a student who's here for short-term training and hopefully help them change their whole worldview so that lifetime learning is now part of their worldview, and that even after they get into the workplace they know that there's someplace they can come and continue to work on their education.

These students endeavor to utilize the services and opportunities around them in an effort to progress. A human services student notes, "I don't think I could make it without [my Pell Grant]. And then EOPS [program for first-generation students], they provide my books and [other] services. I try to use as many of the programs as possible." Several of these students found college graduates who had been in the same program and were willing to give them advice. This is highlighted by Michael, an administration of justice student.

Yes, grads. They introduce me to people that can help me and [those] people . . . help introduce me to more people. . . . It's just a variety of people that I know around here. So, I get a lot of help . . . I never can say I don't get help here. . . . [The staff] help me with my class schedules, help me to fill out financial aid and giving, help . . . me get books, get . . . my whole educational plan.

Faculty and staff are keenly aware of students' needs and the avenues of assistance for them.

The [student success lab] is a place where students can come and very personally and very privately build their skills. (Stephanie, a learning center faculty member)

[We] help students who are low income and essentially underprepared and primarily at that particular point, first-generation college students. And we provide academic advising, personal counseling, career counseling—just about whatever walks in the door with the student. (Vera, counselor for underprepared students, Bakersfield College)

Common to many of these students is their involvement with an organization or department through their work at the institution. Such a connection might be deemed to be consistent with theories of student integration.[8] Outside of the classroom, student involvement with the institution is high. Ellen, a human services student, exemplifies this theme through her comments.

I started off just as a student taking regular classes. I got involved with the federal work-study program. I started out in the work experience program. They kept asking me back. . . . I did orientation, [to] help students. Then I went over to job placement, which is also through federal work-study. I then came and spoke with Manuel [the program director]. I came to a peer/mentor appointment as a matter of fact and my peer mentor, I end up talking to him and he was like, "Wow, you should do this." . . . Now, I am here.

Student actions become consistent with goals when students have the ability and motivation to pursue an appropriate path that is regulated by guides, advisors, and teachers. Guillean, a twenty-three-year-old nursing student whose native language is not English and who was raised in Mexico until she came to the United States to finish high school, notes the importance of her teachers.

My med surg [medical surgery] instructor last semester is very, very helpful. . . . [A]s soon as she'd see my facial expressions—or any students'—she knew we're stressing out or she'd come to us and talk to us and say, "What's going on, relax, take a deep breath, it's going to be okay, what do you need help with?" She was always there to ask us what we need[ed] help with. As well as this semester, we have two instructors that are teaching the med surg class and they're very helpful as well. I know Nancy . . . [one of our instructors] . . . is available for us. She says "anytime. If you guys need me, just call me, and we can get together anywhere. If you're at work and you want me to come to your work, I'll go to your work." On her days off, if we need her she'll meet us here at lab.

When students with lesser academic abilities challenge themselves, as does Ellen, they have both support from the institution and a high level of motivation because their long-term goals are clear to them. Ellen's path to transfer to a university program in drug rehabilitation is lengthy, but the courses that she requires are laid out clearly for her. Even though she does not pass her math class, that does not mean she is deterred in her goal.

I completed Math 50. I am doing . . . Math A, pre-algebra, right now, then I have to complete . . . another math [class], which is intermediate algebra. I have to take one more . . . and it is such a trip because I had Math 50, which is some pre-algebra. I did that online, got an A. You do the assessment and then you go in, [and it is] the same thing you are working on at home on your computer; you just do it in front of the instructor and you get your grade. I was like, "OK, here it goes." I get in there and I just aced it and got an A. And I was like, "Alright, you did this in like two weeks, Ellen. You still have a whole semester to go, why don't you add Math A and finish that one, too." So I was, like, big-headed: "Ya just give it to me, give me all that." Bombed it, I mean, totally bombed out. So . . . I know I am going to have to retake that. I have already talked to the instructor and let her know.

The Role of the College: Wake Technical Community College and Harry Truman College

The students at Wake and Truman articulate an unrealistic understanding of the personal benefits of education. Most of the students think that education will supply them with a better job and more money.

I want to go to the college, too. So, I want it for myself, for something . . . better than the factory. . . . [Y]ou go to the college for something easy and fast and they pay you more and you don't have to work a lot. . . . First, I want to do something so they pay me more, so I don't have to work so hard. (Rafael, ESL student, Truman College)

Students alluded to the view that a good job will immediately follow the attainment of their education: They assumed that employment would be available to them. Harry, an Adult High School student at Wake, noted, "When you start a new job, [you] want to stick with the job at least for a year. Once you have been there for a year, whatever the position you have been doing, unless you have been promoted, you are an expert at it."

Students have been socialized to accept their isolated situation at a site where they are removed from the mainstream college programs and where their naïve views are not disconfirmed by more enlightened interactions with other students. Kelly, a GED student, reflects on her level of comfort at the satellite campus: "I am more comfortable here. It is a lot easier around here. I feel like there are more people my age or older, and I don't feel uncomfortable." Rufus, another Adult High School student, in a series of statements concurs: "I mean, you realize how much you can really learn, you know, once you get past the distractions, the jokes, and the friends and it is just you. . . . I say one productive day here is like a week at a regular high school. . . . It is a good place to go, if you want a second chance." Todd, a GED student, comments, "I now know the importance of education and how it can affect your life when you are an adult. When you are a teenager you don't have a clue." Todd continues, "I think anybody that . . . doesn't have a high school diploma should get that; if they have the opportunity [they] should do it. I think if you put your mind to it you can do it."

A faculty member at Truman, Tom, explains the students' plight:

> For a lot of my students it is like, "If I want to break out of eight dollars an hour jobs, I have got to have a better background and more skills and stuff like that." For them, there is a connection: It is a bridge out of eight dollars an hour, going nowhere jobs that don't pay the rent. "I have got to have something better, and society tells me that the only way I can do it is if I get an education."

Thus, the community college can play an important role in the lives of these students. It may not be the instant solution that some of them are seeking, but it can create a measurable improvement in their life circumstances. Yet, consistent with scholarly work on the limiting results of community college education,[9] these students are fettered in their future opportunities. They see the education and job connection, but beyond that, they see little.

Highly educated immigrants without English language proficiency present a complex twist because they are considered uneducated in the United States until they can speak English. Although they have been attorneys, teachers, or journalists in their home countries, they understand that their inability to speak English limits what they can accomplish in this country. Rayasam, an ESL student at Wake, notes, "My problem was that I couldn't speak English or understand [it], and now I speak English and understand very well. . . . You must learn to speak English. I can speak many things, but my handicap is English." Olivia, another ESL student, comments, "Portuguese is my native language. I speak Spanish also, but here I don't have a job yet. I must learn English first." The ESL students realize that the language barrier is a common problem to overcome to survive and flourish in the U.S. economy and society, and the community college plays an important role in assisting them.

Those in ESL who were more educated were able to articulate both their ideas and aspirations. These articulations were rare among the other less well-educated students, suggesting that the prior education of these ESL students contributes to their understandings of the value of college. Tia, an ESL student, notes, "I am not in a position to choose what I want to do right now. There is a lot involved. You know, like money matters or disabilities, you know, a lot of things come. I have to listen to reality right now and do what it presented to me." Tia continues, "It is really hard, especially when a lot of people on the street you can't talk with them on the street as they do. They think you are stupid. It is hard, because you can't explain that to them. Of course you can; they just won't understand you."

Surprisingly, the students do not see themselves as pushed to go on for further education or into specific jobs. The faculty and staff seem only to help the students' progress toward the goals that the students have for themselves. The faculty and staff do not exert influence on the students. Many of the students do not even know all the types of classes offered, the available services, or the flexibility in class times. Students expressed a need to be motivated by others. Cindy, an ESL student at Truman, states, "I am kind of lazy and I need somebody to push me." Another ESL student, Randall, concurs: "I have been lazy. All my life even in high school, you know: lot of trouble." Tia, also an ESL student, remarks, "No, we haven't spoken with any advisors. If I get a chance to do it I would." Antone, an ESL student, notes, "In here, the people around here tell me that the money is not important, but I have to be honest with [myself]. If I want to live in here, I have to have money; I can't afford to do what I really want to do if I don't have the money." Tatiana, a GED student, comments, "I should study hard for myself. [But I would like] something that would give me more pressure to study. Okay, you have got to read all of this and tell me tomorrow. I would like some homework, but they said 'no.'" Students are neither motivated nor directed by the institution.

The faculty and staff think that the role of the community college is to reach these students who would not succeed at another institution or agency. But the definition of success is both educationally and economically limited. The faculty and staff articulate their efforts to help students obtain a better job and complete the academic program or class in which they are enrolled. But the faculty and administrators also characterize the instructional role narrowly, limiting their ability to give the students the knowledge and skills needed in the current job market.

> Our objective is to try to get them into a track to find a job and succeed and get a better paying job, besides being a waiter, or a busboy, or a dishwasher. . . . Of course, we focus on the four learning skills needed by the workforce. (Dean of Basic Education, Truman College)

While there is a plan to make alterations, they have not yet been integrated into the classroom. A faculty member, Tom, explains: "They plan to significantly change the curriculum so that instead of the English programs, teaching English acquisition, they are going to be more narrowly focused on particular job areas and particular industries." Tom discusses the special needs of this population and what he is required to do to assist them.

> We can have discussion on world events or psychological counseling and how it has hurt or helped the people in the room, or child abuse and all kinds of things, but at the end of the day I have to tie it down to preparing them for a test. . . . They came here to prepare. Mainly, it is to prepare them for the test. The test has changed and there is a lot less emphasis on reading comprehension and a lot more on critical thinking skills, inferential reasoning skills, trying to find out the assumptions, being able to find out which conclusions are based on the evidence presented in front of them. I hope I have adjusted my teaching to that. That is mostly it: [the focus upon] the outside world. Computers are more important, also. We go to computers once a week now.

Students at both Wake Technical Community College and Harry Truman College are in programs that are customarily classified as remedial or developmental education. Either previous academic achievement (or lack thereof) or present academic deficiencies are characteristics of these populations. One large group of students did not complete precollege coursework in secondary school. A second group includes immigrants who lack English language skills as a result of the absence of English in their formative years. For the English as a second language students, there are two major categories—those with academic skills gained in their home country and those who had little education in their country of origin. This latter group exhibits characteristics similar to those non-ESL students who are in need of remediation or development or both in order to cope with coursework. This population, the academically underdeveloped—whether ESL or non-ESL—is vulnerable on several counts. First, they are "at risk" for failure to complete coursework. Second, they are in jeopardy of facing a lifetime of underemployment or poverty-level employment. Third, they experience their education through personally underdeveloped identities, as peripheral to social and economic life in their communities—indeed, they are students beyond the margins.

The Role of the College: Bakersfield College

Through the structure of Bakersfield College, in some distinction to Wake and Truman colleges, students have developed a coherent and meaningful sense of direction. Eduardo, an anthropology and forestry student, notes, "[The institution] has helped me a lot. As far as the direction that I wanted to go into, it kind of took time, but it formed the direction that I'm in right now." Yet to achieve their goals they must be self-motivated. An applied science and technology student, Kathleen, states, "The students start college and find out it is not like high school; you don't have to go, actually; nothing really bad happens if you don't go, except for at the end when you don't pass the course."

These students want to obtain education beyond the community college. Karen, a reentry student, notes this: "I will go on to get my BA and my master's, which I never, when I first got here, I would've never believed possible or had never considered, because it wasn't in my realm of possibilities at the time." Ellen, a human services student, remarks, "I plan to transfer after I am done here. This last year I have really been cramming my classes and taking everything I need so I can get out of here." A political science student, Grace, notes, "I want to finish my general [requirements] here. After that, I want to transfer, where I am planning to get a BA." Maria, an applied science and technology student, comments, "In two years I graduated with my AS. . . . I am to get my master's. That I have already mapped out. It is just five classes . . . after I get my license." An ESL and prenursing student, Nidia, states, "My thinking is, later on, get my master's." James, a liberal arts student, remarks, "I can't go to a university because it is too expensive . . . but I will transfer to a college at the end of the semester."

These are students not only with future expectations, but also with an understanding about how to fulfill those expectations. Their designated identities combine how they are performing in the present with realistic expectations about their future goals and the actions required to attain these targets. The role of the college has been in part to assist them in both framing their goals and findings ways to attain them.

The Three Colleges Compared

The examination of three colleges—two of which are featured here primarily through their Basic Skills and English as a second language students and one that is featured through a broader spectrum of program students, but including students in developmental and ESL courses—indicates that the framing and contextualizing of student education have salience for expected student outcomes. At Bakersfield College, the students featured here, while possessing many of the identity characteristics of the students at Harry Truman College and Wake Technical Community College, appear to have greater potential for college completion and gainful employment than those students at the other two colleges. While all the students fall under the category of "nontraditional students" and most have several nontraditional characteristics, the Bakersfield students exhibit a pattern of educational direction and steady progression through their program goals to further education or meaningful employment. This is not the case for the Truman or Wake students. A number of factors contribute to this pattern at Bakersfield, not present at Truman or Wake.

Even though the students at all three community colleges had received assistance from faculty and staff to reach their academic goals, the students at Wake and Truman College were rarely encouraged to look into fields that were in high demand or to seek further education. The students at Bakersfield College were encouraged to continue their education beyond the community college and to look at other fields or careers they may not have known existed or were in high demand for employees.

While it was noted that the support services and staff were available for all the students, all three community colleges were experiencing budget cuts that were diminishing the level of services available to these students. Each institution had varying types and levels of services available to their students, yet a difference noted at Bakersfield College was the higher rate at which students utilized support staff and services. Nearly every student interviewed at Bakersfield College utilized financial assistance programs, worked with support staff for guidance and direction, and developed strong relationships with faculty, staff, and other students. These relationships and support services were more of an anomaly with the students at Wake and Truman. The students at Bakersfield College understood the value of the relationships they are developing. Bakersfield College students were also highly active and engaged with the institution's community, which was not the case at the other two institutions. The students at Wake Technical Community College and Harry Truman College were often unaware that support services and staff existed, and those students who did know the services were available rarely used them. Textbooks and technology were also often outdated for the students at Wake and Truman, whereas the students at Bakersfield College purchased their own books and did not voice concerns about the available technology.

At Wake and Truman, students were not challenged academically, and instructional methods were aimed at satisfying basic needs. What is referred to as constructivist learning[10] was not evident at either Wake or Truman. Students were not part of the directing of the learning process; rather, they were the objects of learning. Students at Wake and Truman were not given homework and did not work in groups, while the students at Bakersfield College were often given homework and worked in groups on a regular basis.

The students at Bakersfield College exhibited knowledge of social and economic matters, while the students at Wake Technical and Harry Truman College demonstrated relatively limited knowledge of these areas. This lack of knowledge has a bearing on students' understanding of both their educational and occupational goals.

The students at Wake and Truman are enrolled in classes, but they are uncertain of what classes they need to take to reach their academic goals. Even with academic goals, many of the students do not know what they will be qualified to do in the workforce once they complete their program. These students often have unclear goals and act with little vision or direction. The students at Bakersfield College, in contrast, are focused on and familiar with the classes they need to take and the degrees they intend to obtain. They know what they expect to accomplish and where they plan to work when they are finished.

The students at Wake and Truman are also personally or individually oriented. They are not concerned about their role in society and their actions and goals are driven by their personal concept of happiness. The students at

Bakersfield College are focused on society, others, and their family, and they react accordingly.

The students at all three sites experience considerable diversity of backgrounds and characteristics in their classes. While the students at all three sites acknowledge these differences, the students at Wake and Truman did not express a bond with their fellow students. The students at Wake Technical Community College understand they are all seeking a similar education, but perceive that their actions stem from different motivations. The students at Harry Truman College were aware that they share similar academic goals and appreciated what they could teach one another, but they did not develop a strong connection with their fellow students. In some distinction, Bakersfield College students were able to see that different people were in the same program together trying to complete an academic program. These students expressed their feelings of a common bond, looked out for one another, and tried to ensure they all passed each class.

The students at Wake and Truman seek an education to meet their basic needs. In general, they want only enough education to be qualified for a job that will allow them to live a "happy" and comfortable life. Many of the students did not desire education beyond the community college, and they assumed that the Basic Skills (Adult High School, ESL, or GED) they were receiving would allow them to find a job where they could remain financially solvent. These students were unrealistic about the benefits they would receive from completing these Basic Skills programs. Although Bakersfield College students know a certificate or an associate's degree will help them attain a secure and well-paying job, this will not solve all their problems. The students know that there will be more obstacles to overcome and that they will likely need more postsecondary education. The students at Bakersfield College desired to obtain further education beyond the community college.

Wake and Truman students are also unaware of the consequences of not obtaining any further education. The students were oblivious to their future in the workforce and their financial well-being, with or without an education. The students at Bakersfield College understood what failure on their part would mean for their financial future. These students were willing to take risks and give extra effort to ensure they would reach their academic and work-related goals. The students at Bakersfield College accept incurring some debt to achieve their academic goals, whereas many of the students at Wake and Truman stated they would not seek further education if it cost more.

The students at Wake and Truman did not actively seek advice or suggestions about their future, but if given suggestions on how to reach their goals, they often stated interest in accepting the advice. Their lack of awareness made every proposition seem valid and, to them, almost "too good to be true." However, these students rarely acted on the suggestions they were given. Bakersfield College students often sought advice and took suggestions, but made their own decisions after weighing all options thoroughly. Bakersfield students were involved and active in their educational decisions.

Conclusion

The question is, therefore, Are students at all three colleges receiving just treatment? Do college contexts make a difference? The literature on the community college generally takes several distinct positions on the effects of college on students. One position is clearly that the effects are positive,[11] that without the community college millions of students would have no postsecondary educational experience.[12] A second is more circumspect, although it tends to side with the positive camp and indicates that there are a number of issues, such as weak strategies, that need to be improved if students are either to acquire knowledge or gain in their station in life.[13] A third is critical of the institution for failing to make substantial gains either in student learning or in student mobility.[14] This chapter indicates that outcomes are contingent upon institutional factors, and that some institutions have positive effects upon student learning and ultimately upon student mobility or their station in life. In this chapter I focused upon three community colleges and compared student behaviors and institutional actions with respect to students, and then identified both outcomes and potential outcomes for students. Although this is a comparison of three institutions, the differences between one institution's outcomes and those of the other two indicate that institutional context has a bearing upon the treatment of students. In some respects, these students at Bakersfield College, Truman College, and Wake Technical Community College are all disadvantaged students with academic deficiencies compared to traditional college students, as well as behavior patterns, such as their dropping out of school or college, or life experiences, such as substance abuse or illness, that have conspired in the past to frustrate their participation in postsecondary education. Furthermore, students whose native language is not English or students with disabilities have additional disadvantages. College attendance by itself is not according students justice in the sense that John Rawls[15] explains justice: These students are advantaged only if they gain from their college experience—that is, if they overcome their disadvantages educationally, socially, and economically. Clearly, the Bakersfield College students noted in this chapter have gained as they progress in their coursework, in their programs, and in their further education at university or in their careers. In addition, they have gained in their knowledge of themselves and their society. The evidence for Truman and Wake students is less convincing.

Notes

1. I am indebted to Jerrid Freeman for this comparison, as Jerrid worked for me as a research assistant in 2003–4 and subsequently used my sites and the data from these sites for his doctoral dissertation.
2. Kathleen Shaw, "Defining the Self: Construction of Identity in Community College Students," in *Community Colleges as Cultural Texts*, ed. Shaw, James Valadez, and Robert Rhoads, 153–71 (Buffalo: State University of New York Press, 1999).

3. Anna Sfard and Anna Prusak, "Telling Identities: In Search of an Analytic Tool for Investigating Learning as a Culturally Shaped Activity," *Educational Researcher* 34, no. 4 (2005): 14–22.

4. There is a considerable body of scholarship on student outcomes that combines student characteristics and institutional effects. Estela Bensimon, Linda Hagedorn, and Tom Bailey are among those scholars who have made the most significant inroads on community colleges in the past decade. See as examples: Thomas R. Bailey et al., *Improving Student Attainment in Community Colleges: Institutional Characteristics and Policies* (New York: Community College Research Center, Teachers College, Columbia University, 2004); Linda Hagedorn, *Traveling Successfully on the Community College Pathway: The Research and Findings of the Transfer and Retention of Urban Community College Students Project* (Los Angeles: University of Southern California, Rossier School of Education, 2006).

5. For Wake Technical Community College and Truman College, I use pseudonyms for students. The maintenance of anonymity of individuals was a condition of conducting interviews with students on these sites; at Bakersfield College, there were no conditions, thus the names used are actual ones.

6. See college profiles in appendix.

7. W. Norton Grubb, *Honored but Invisible: An Inside Look at Teaching in Community Colleges* (New York: Routledge, 1999); Vincent Tinto, Anne Goodsell Love, and Pat Russo, *Building Learning Communities for New College Students* (State College, PA: National Center on Postsecondary Teaching, Learning, and Assessment, 1994).

8. Ernest T. Pascarella and Patrick Terenzini, *How College Affects Students: A Third Decade of Research* (San Francisco: Jossey Bass, 2005).

9. Steven Brint, "Few Remaining Dreams: Community Colleges since 1985," *The ANNALS of the American Academy of Political and Social Sciences*(March 2003): 16–37; Kevin Dougherty, *The Contradictory College* (Albany: State University of New York Press, 1994); James Valadez, "Cultural Capital and Its Impact on the Aspirations of Nontraditional Community College Students," *Community College Review* 21, no. 3 (1996): 30–44.

10. Grubb, *Honored but Invisible*.

11. Terry O'Banion and Associates, *Teaching and Learning in the Community College* (Washington, DC: Community College Press, 1995).

12. Arthur Cohen and Florence Brawer, *The American Community College* (San Francisco: Jossey-Bass, 2003).

13. John E. Roueche, Eileen E. Ely, and Suanne D. Roueche, *In Pursuit of Excellence: The Community College of Denver* (Washington, DC: Community College Press, 2001); Roueche and Roueche, *Between a Rock and a Hard Place: The At-Risk Student in the Open-Door College* (Washington DC: American Association of Community Colleges, 1993); Roueche and Roueche, *High Stakes, High Performance: Making Remedial Education Work* (Washington, DC: Community College Press, 1999).

14. Brint, "Few Remaining Dreams"; Dougherty, *The Contradictory College*; Grubb, *Honored but Invisible*; David Labaree, *How to Succeed in School without Really Learning* (New Haven, CT: Yale University Press, 1997); Rhoads and Valadez, *Democracy, Multiculturalism, and the Community College* (New York: Garland, 1996); Shaw, Rhoads, and Valadez, eds., *Community Colleges as Cultural Texts* (Albany: State University of New York Press, 1999).

15. John Rawls, *A Theory of Justice* (Cambridge, MA: Belknap Press of Harvard University Press, 1999).

Chapter 6

Strategies and Actions of Community Colleges, and Their Behaviors, in Accommodating Nontraditional Students

Introduction

Universities and four-year colleges in the United States have traditionally served as substitute parents or as social and intellectual guides, or both, for recently graduated high school students. In the community college, in part because of its historical development and in part because of its adult student population, institutional focus is less on the social and intellectual development role or the "rearing" of students and more on social services, educational and vocational advisement, and recently, on learning processes and outcomes.[1] On the one hand, with its self-proclaimed identity as a teaching and student-centered institution, the community college has increasingly addressed student outcomes, including further education and employment and learning. These outcomes serve as markers of student mobility, economically as well as socially, and reflect the community college's position as part of what scholars view as its place in the educational social structure.[2] On the other hand, and in large part consistent with state funding based upon student enrollments, community colleges have focused upon the recruitment and retention of students. Indeed, retention of students—their continuation beyond one course or one semester or one year of enrollment—has become a major concern of the institution, guiding institutional strategies that frame understandings of students, particularly nontraditional students.

While strategies of the institution are oriented toward student retention as well as student performance—strategies that have become synonymous with institutional goals—behaviors are less organized around principles and initiatives.

Behaviors associated with the community college are a conglomerate of individual behaviors of students themselves as well as administrators, faculty, and staff. Behaviors of students, or directed at students, are comprised of teaching and learning; other behaviors are interpersonal interactions, relationships between students or between students and faculty, and the actions of students themselves in their decision making about their course and program choices, and their work within those activities. A substantial portion of these individual behaviors are aimed at student mobility—such as access to the institution, movement of students through courses and programs, and movement beyond the institution to further education and employment. These patterns of progression can also be viewed as social mobility, but that view might understate the effects of college upon nontraditional students.

Is It All about Social Mobility?

One of the major arguments in higher education over the past five decades—an argument specifically focusing upon the community college—concerns the social mobility effects of college upon students. While the promise of community colleges has centered upon individual advancement, both economically and socially,[3] scholars from Burton Clark to Kevin Dougherty have challenged the claim.[4] In the past decade, the challenges have withered, and to some extent they have been addressed by recent scholarship that provides different analytical methods and employs other theoretical perspectives.[5] Yet, the issue is pertinent for two reasons: first, because practitioners continue to make the claim that the community college advances the prospects of students, and second, because without social mobility for students who are disadvantaged, the institution serves as a place of containment for a large segment of the population and these students are treated unjustly.

My investigations indicate that the answer to the question of social mobility for disadvantaged students is mixed, complex, and nuanced if we examine the views of community college practitioners on this question. Even an ardent supporter of community college education for students, the president of Edmonds Community College, indicates that the issue of social mobility is muddied by the problem of resources, as his college is not gaining enough revenues from the state to support student education. Yet, without the community college, the reverse of social mobility—a process or condition of social degradation—is likely the outcome for large populations.

> I think education is . . . social mobility and . . . [the] community college is certainly for lower-income folks, for persons of color, and for first-generation college students [and] is often the last place that's available. . . . [I]n my view . . . you're shooting yourself in the foot by serving all these people if you're not getting paid for them. But if we don't let them in, where are they going to go? Where are they going to go? Well, they're going to go to jail; they're going to go on social [welfare] roles; they're going to go

to the local bar. They're sure not going to be contributing to society. (President, Edmonds Community College)

There is a wide range of understandings within the community college of what constitutes social mobility, as noted in a number of responses from institutional members—administrators and faculty, as well as system executives—in nine states covered by my investigation.

> So we've got corporate and community services, small business development centers. The students that are served through the contracts are all employed by the employer, so if they move on it's because their employer values education, supports them moving on, sees the value of going ahead and taking your English class, your math class, your traditional core general education, and getting an Associate of Arts degree. BA (A)9

> I definitely think [college contributes to their social mobility] . . . along with whatever structure we provide in the program because we almost train them that they have to do certain things, rather than just letting things do whatever they want. BA (A)1

> Students drifted to us, didn't have any clue why they were there. They had some vague idea of improving their lives and really then made a giant step up in terms of who they were and what they were able to do with their lives. BA (A)25

> I hadn't looked to that as an outcome, but I would think it's [social mobility] there. . . . [W]e are preparing them to be learners and earners. BA (A)27

> I think that it adds to the overall self-esteem that they are self-supportive. Some people, unfortunately, come to nursing school so that they can be self-supportive, so they can divorce. And not have to be dependent on somebody else. Some come back from the opposite reason: They got a divorce and they had no way to support themselves. BA (F)14

> I want them to believe in themselves, that they can improve their lot in life, absolutely. BA (F)16

> [W]e want to make them successful citizens to improve their lives. BA (F)30.

> I would say economic [mobility]. . . . [H]ere in New York City, a number of our students come from communities where they're not expected to make great achievements. And so when they come in and you work with them to, one, establish goals and, two, work toward those goals and ask they progress through, they can see the reality, or at least sense these goals are achievable, and they can themselves then be placed in a certain setting later on in life. There is a transformation and, yes, they do move from here, or one social status into a totally different status when they leave. . . . [T]hey do move from here on to paying position jobs that they had not . . . thought that they could place themselves. So, yes, it really does provide the platform for them to move on and do great things. BM (A)2

I think that . . . students don't see themselves as second-class citizens. They see themselves as coming in here, getting the first two years and then moving on. BM (A)3

I don't see any better doorway to the first rung on that social ladder than the community colleges. . . . The community college is the perfect place to help open the door and move them up the food chain. CA (E)2

Amazing diversity in California, with lots of new blood coming in all the time, in terms of immigration and amazing levels of haves and have-nots. I don't think that mix would work if you didn't have something that provided readily accessible social and economic opportunity. I don't know how the equation would work. CA (E)1

I think in every aspect of their lives there is social mobility for community college students: communication, security, finding a job if they don't have one. Many of them get promoted. DA (F)6

[Y]ou can't undo what's been done for all their lifetime, but many of them walk out with . . . a better use of the English language, a better idea. DE (A)1

Any time a student can get more education and improve their world-view and their earning power they win. They benefit. DE (A)10

Community college students come here for very different reasons than to graduate. Most are not here to graduate from this institution. Most are here to gain skills to go to work, most are here to gain enough credits to move on to the [university], most are here because they're in a job that requires upgrading, so in that regard all three of those outcomes are things that will improve their economic futures. PI (A)5

We are trying to help them be successful. . . . Our objective is to try to get them into a track to find a job and succeed and get a better paying job, besides being a waiter, or a busboy, or a dishwasher. TR (F)12

Table 6.1   Social Mobility

Codes
College: Bakersfield College (BA), Borough of Manhattan Community College (BM), Community College of Denver (DE), Pima College (PI), Dallas Community College District (DA), California Community Colleges (CA)
Position: Policy Executive (E), College Administrator (A), Faculty (F)
Individuals within Category: 1, 2, 3, 4, 5, 6, 7. . . .

To the question, "Does the community college contribute to student social mobility?" college administrators and faculty responded in the affirmative, although their views contained mixed messages and the concept of mobility was more often than not economic. Indeed, several of the administrators and faculty viewed their college as a pipeline to employment.

Oh, yeah, that, that I definitely think so, yeah, along with whatever structure we provide in the program because we almost train them that they have to do certain things rather than just letting them do whatever they want. So . . . one of the things that we're going to institute for next year, we . . . [will] require everyone that participates in our program, require them up-front to develop a résumé. So that they, if nothing else, when they leave here, they have a résumé and maybe even know how to interview. (Administrator, Bakersfield College)

Indeed, the role of employment and employers has considerable salience in the community college, especially in the continuing education and contract training units. In these program areas, student mobility may be dependent upon employers' views of the benefits of college education.

So we've got corporate and community services, small business development centers. The students that are served through the contracts are all employed by the employer. So if they move on it's because their employer values education, supports them moving on, sees the value of going ahead and taking your English class, your math class, your traditional core general education, and getting an Associate of Arts degree. (Administrator, Bakersfield College)

In fields, such as nursing, mobility is seen in economic terms, and the associate credential in nursing has considerable cache in the labor market.

Nurses are paid hourly, so they'll start at $20 to $25 an hour. Bakersfield's on the low end. You go LA, San Francisco, they're starting at $40, $45 an hour. And that's with a two-year degree. So definitely [there is mobility], especially when we've had welfare-to-work students that come in and are successful. (Nursing faculty, Bakersfield College)

Yet, this same nursing faculty indicates that simply wages are not the whole story for the economic mobility of her program: that economic self-sufficiency leads to social benefits and the release of personal potential in the realm of individual accomplishment.

I think that it adds to the overall self-esteem that they are self-supportive. Some people, unfortunately, come to nursing school so that they can be self-supportive, so they can divorce and not have to be dependent on somebody else. Some come back from the opposite reason: They got a divorce and they had no way to support themselves. So there's that contribution that I think you see to society . . . because of your earning power and what you can now do, and you never thought this was possible and the accomplishment that happens and occurs by being successful. (Nursing faculty, Bakersfield College)

Others view the mobility issue as, on the one hand, fitting in to the mainstream of American society and, on the other hand, as personal development and advancement.

> So we provide those three component areas of support to help them become mainstream, because the notion has been that a number of these inner-city students were not initially able to go to college.
>
> A number of our students come from communities where they're not expected to make great achievements. And so when they come in and you work with them to, one, establish goals and, two, work toward those goals and . . . they can see the reality or at least sense these goals are achievable and they can themselves then be placed in a certain setting later on in life. There is a transformation and, yes, they do move from here, or [from] one social status into a totally different status when they leave. Because then they move on. . . . And, yes, they do move from here on to paying position jobs that they had not [previously] thought . . . [where] they could place themselves. So, yes, it really does provide the platform for them to move on and do great things. (Director, Discovery Program, Borough of Manhattan Community College)

Personal enrichment and development are particularly evident in students who have come to the community college from disadvantaged backgrounds: for example, those who are second language speakers of English.

> I think in every aspect of their lives: communication, security, finding a job if they don't have one. Many of them get promoted. It's very exciting when they come back and tell you, "Teacher, thank you. I learned more English. I got more money." . . . [There is] more involvement with their children in the schools, which is neat also because they get comfortable enough to be able to attend PTA and the programs their kids are in, talk to the teacher and different things: helping their families, that's a big one; learn English, work here; send money home; bring families here— that kind of thing. (Faculty, Mountain View College)

A program administrator at the Community College of Denver notes the role of a college strategy in developing mobility for students.

> Yes, the soft skills [are] a large piece of what we do. And now, granted, you can't undo what's been done for all their lifetime, but many of them walk out with a better use of the English language, a better idea. We used to teach a course, it used to be called personal growth and development, and in that course we talk about hygiene; we talk about [how] you have to show up on time: you have to make a commitment to be there five days a week. We watch attendance very strictly because we're trying to get them into the habit of coming on time and leaving on time like they would on

a job. . . . [On] Wednesdays we try to do 'dress for success' and encourage them to dress as you would on a job so that they start looking at the wardrobe and what they've got, and encourage them to do so. (Program Administrator, Community College of Denver)

The answer to the question of social mobility for community college students is not a straightforward one. On one level, the acquisition of skills and knowledge is going to enhance a student's employment opportunities and social awareness, as a dean at the Community College of Denver notes: "Any time a student can get more education and improve their worldview and their earning power they win. They benefit." On another level, the outcomes of community college education— even the outcomes after the gains of skills and knowledge—are limiting. "We are sometimes too reactive to the business community and build programs to fill a pipeline and put people into jobs that maybe they didn't want, weren't interested in, or that are dead ends," notes another dean at the same institution. The social mobility issue is especially evident when students who are not advantaged—who lack appropriate academic backgrounds for college or are not native English language speakers—are the object of an institution's goals to train students, as noted by an instructor at Harry Truman College in Illinois: "Our objective is to try to get them into a track to find a job and succeed and get a better paying job, besides being a waiter, or a busboy, or a dishwasher."

The question of social mobility is complex. Although Arthur Cohen and Florence Brawer argue that students do learn, become employable, and are able to move between institutions—certainly a form of social mobility— and that there is no alternative to the community college,[6] our concerns should not stop there if John Rawls' "difference principle" is to be accepted. This principle underpins justice as fairness: "[S]ocial and economic inequalities . . . are to be adjusted so that, whatever the level of those inequalities, whether great or small, they are to be the greatest benefit of the least advantaged members of society."[7] For the institution called the community college, the question is not so much whether students achieve social mobility, but does the institution adjust social and economic inequalities for its students?

Institutional Strategies and Actions

Strategies and actions within colleges focus upon assisting students to access the institution and its programs, supporting students in their courses and programs, improving student academic performance, enhancing student experiences, and retaining students at the college or facilitating their movement to further education or employment. In California at Bakersfield College, in the developmental learning program area, the use of Learning Communities for remedial education has proved to be one of the more successful approaches for underprepared students, who are predominantly Latino, with African Americans as the secondary most populous group in the program area, according to the faculty chair of developmental education at Bakersfield College ("A" signifies the interviewed person and "Q" signifies the interviewer).

A: Right now we do have some Learning Communities going, too—Developmental Learning Communities—which are interesting, so that [students] would at least have exposure to people in different disciplines. . . . The focus of the one in the fall [semester], and actually . . . we're focusing on it this semester, it's kind of a career search, career exploration. So first semester . . . they develop their education plan and those kinds of things, and now they're really kind of delving into a particular career. . . . We've got a project formulated whereby they will . . . write the paper for the English class. They're gathering information for the counseling class, and they'll present a speech in my class. . . . We're finding this to be very successful. . . . A number of students . . . have gone and started this particular Learning Community in fall. Now they're together in spring, and they'll continue one more time in the fall. . . . One of the connections we've had is, we've kind of loaded them up to do their full schedule each semester. . . . Basic skills research shows that it's very important to get students to complete as quickly as they can: I mean fifteen or thirty units within a year. Their chances of success are much higher. So we're trying to make sure that we provide that for them. . . .

Q: Why do you call it a Learning Community?

A: Because, we try to integrate the curriculum as much as possible with disciplines. . . .

Q: Multidisciplinary?

A: Yes, it's multidisciplinary. What's been interesting about this Learning Community, though, is that, I want to say it's almost, it's lower in terms of the students' abilities when they come in than a regular typical class. For example, I would have a typical writing class. . . . I would have some people high, some people low. This Learning Community, they're all low because they're low in writing, and they're low in reading, and they're low in math.

Q: How do they get there?

A: We spoke with the counselors and if a student fits that particular profile, then that counselor would recommend: "Well, here's a program that might be for you." And overall I would say that success rate has been higher than for students who would have self-selected those classes.

The above approach is consistent with institutional strategies from a trait approach to improve student achievement and particularly the retention of students.[8]

Other strategies for nontraditional students include specific programs for subpopulations, such as first-generation programs for students who are the first in their families to attend postsecondary education and adult high school programs for students who want to complete a high school diploma on a college campus or who attend a high school at a college campus and take college-level courses either concurrently or subsequent to their high school courses. First-generation programs, using federal funding, target populations that are merely samples of the larger first-generation population on campus. They serve as programs that advantage students, giving them not only services—such as tutoring—but also privileges—such as course registration priority. Of particular importance is the role these programs play in student persistence.

Karla at GateWay Community College in Arizona attends the college's affiliated charter high school and takes her prerequisite classes for the nursing program. Her participation in this adult high school program includes free tuition and books. Karla is not a native English speaker, and thus neither college nor high school is a simple matter.

> The first day I had a headache. I knew a little bit, like I knew some English but not that much and I had a headache. I was like, "Oh, my gosh." It just sounds like a lot of bees talking and I was like, "No, stop." But then I started to get used to [it]. . . . [E]verything has gone really well. I enjoy my nursing classes. I love them, the ones I already start. And my high school classes, I guess all of them. My math classes have been really nice. (Karla, GateWay Community College)

Derek, who did not fare well at high school, entered the adult high school program at Edmonds Community College, where he entered a program funded by the state. He completed his high school program and then moved on to his college nursing assistant certificate program and an associate's degree, through which he will have both EKG technician and phlebotomist technician status.

> I went to Jackson High School, Everett school district, and I actually, I wouldn't say I dropped out, but I dropped, kind of left there for a program that they had here. It's kind of like a running start. It's called EDCAP, Edmonds Career Access Program. . . . And what it does is, I get college and high school credit at the same time and I'll get a diploma and not a GED, and when I graduate I'll also have a two-year college degree, which is this spring. (Derek, Edmonds Community College)

Without these strategies—program establishment and the pursuit of funding by these institutions—Karla and Derek would be bypassed. Karla would be stuck in a service-level job, at best, with her lack of English language skills and without academic coursework that will transfer to Arizona State University. Derek would join the Army after finding, if lucky, a way to support himself in a GED program, or else he would simply be a "dropout" and find or lose his way in a service economy, shut out from a knowledge economy and society.

Institutional strategies can be more global, not merely single or limited programs or services, but based upon traditions, past practices, and institutional culture, as evident at the Community College of Denver (CCD), where a rational system of intervention—referred to as case management—pervades the institution.

I think the tradition at CCD has always been to examine how we are doing as an institution and how students are doing. I think that has long been in place about looking at student data. . . . And then we've just broadened that . . . and that means the resources; that means going for more in other places to get the work done that needs to be done. . . . [A]s an Hispanic serving institution, we are eligible and seek very actively every federal grant we can get to serve these student populations, and a Title V grant is a very significant first-generation program, which truly is a remarkable program. . . . Our approach to students services . . . is through case management. . . . [S]tudents don't see . . . whoever is available; they have assigned advisors the entire time. . . . [T]his case manager's job is to make sure that the student knows . . . about all the services. It's almost like coordinating various services for the student. (President, Community College of Denver)

Indeed, the president, and she asserts that this is a common institutional perception, does not view students in categories found in the trait framework, which suggests a deficiency model of students.

The way I think about them, and I think that is true of CCD . . . is that they're not "at risk." I think a lot of people [in other institutions] think of students like this, well, you know, "These are difficult." I think, "No, this is who our market is; these are the students we serve; and they have a lot that we can learn from them about their life experiences." And so . . . they're not at-risk for us. We just know how to work with them so we know what it takes. And I would say that is widely held by the college.

Yet, traits have their place in institutional strategy, largely because of the demographic characteristics of both the student population and of the community: 32 percent of CCD students are Latino; 16 percent African American; and 10 percent Asian. In the words of the president, "You have to know who you serve to make sure you serve them well."

Demographics are definitely one of the lenses which we view . . . because it's important to track the students. . . . "Who are we letting in, and are they succeeding?" And if there are disparities, we would want to know why. . . . When you think of the public school side, there are huge disparities among ethnic groups and huge disparities between men and women. . . . We're heavily women . . . we're 62% female. (President, Community College of Denver)

The Community College of Denver grounds institutional strategy on broad categories of student needs, framing these needs in assumptions about student characteristics—which include ethnicity, economic status, and gender. The goals of strategies such as first-generation programs are to combine student access (i.e., participation in postsecondary education) with student attainment (i.e., course completion, program completion, and the movement to either further education or employment). This pattern is consistent with research reported by John Roueche and Suanne Roueche and Roueche, Ely, and Roueche on CCD in the late 1990s and early 2000s.[9]

Specific populations and their disadvantages lend themselves to particular strategies and actions. At Pima College in Arizona, a program administrator underlines both the significance of his program as well as unequal treatment of a population of students based upon state funding. He articulates the experiences these students in their efforts simply to cope with their conditions and as students at a postsecondary institution.

> We are dealing with, in some cases, some of the most troubled and, on occasion, troubling youth in the city. We don't have as many problems as you might expect, but these are kids that have raps, the K through 12 management usually, because they haven't done well there; they have been asked to leave or they aren't welcome. They are coming back, many of them after being out for a year or two, discovering that things don't look good if you don't have a certain level of education, [and] of course, high school is the minimal standard. Ya, it causes pressures all the time; K through 12 money does not follow them into adult ed. I was just up with the legislature this week and . . . the information I was sharing with them is that in Arizona, we have a $2.8 billion K through 12 budget. The adult education allocation of that budget represents .00142% of that total, $4.4 million. That is what they get for adult education in this state, about one-seventh of 1%. That program, however, supplies over 25% of the high school graduates in Arizona every year through the GED program. . . . The population that we are engaged with is growing by leaps and bounds, and they desperately need gateway opportunities as a people, gateway educational opportunities to improve their lives and the lives of their families. . . . I think that community colleges as an institution are probably going to be their most important resource . . . maybe the most important of all the education, you know, not just for our community, but for communities across the country. (Adult Basic Education program administrator, Pima College)

The unequal treatment, he contends, is a consequence of these adult learners' powerlessness. The college, however, maintains this program in spite of public perceptions of college that ignore this population. Instead of accepting the standard rhetoric that justifies postsecondary education—preparation for the workforce, development of a professional class—this program responds to individual student needs so that these students can cope with the vicissitudes of contemporary life and the tribulations of their own lives. Here, this college administrator combines the perspective of a policy analyst with the values of a social critic. As an administrator, he is hardly a bureaucratic manager. Instead, he acts with personal conviction, and the conviction is founded on a sense of justice for this needy population.

> So, this giving people this opportunity, you know, people look at them, getting an education in lots of different ways. Here with Pima adult education, I think, a lot of people look at AE [Adult Education] programs [and say], "Oh, these are people that prepare people for business," and I like to look at it, rather, as "This is a program that prepares people to do business: to do public business, economic business, educational business." We look at it in a way that is a little bit broader. Do you understand the distinction I am making? (Adult Basic Education program administrator, Pima College)

In addition to their educational disadvantages, these students, as well as those who work with them, must confront negative social attitudes about specific populations, including immigrant populations, welfare-to-work populations, and the working poor. In the 2000 census, there were 28.4 million people living in the United States who had been born elsewhere, or about 10 percent of the total population.[10] Immigrant youth are twice as likely as youth born in the United States to live in low-income families, and they are more likely to have parents with lower levels of educational attainment.[11] The distinction made by the Pima College administrator is one of broadening the goals and outcomes of his program beyond economic development to citizenship and the public good. By educating these students, his program provides them with the skills and knowledge for their participation in society as workers, learners, and citizens.

Welfare mothers are another disadvantaged group that requires particular strategies by colleges to respond to their conditions. Jerry Jacobs and Sarah Winslow and Rebecca London indicate that policy reforms in the late 1990s for welfare have limited the chances of welfare mothers to pursue higher education.[12] They also note that policies differ state to state and that these policies contribute to the treatment of welfare recipients. Mazzeo, Rab, and Eachus concur on this point about policy differentiation state to state.[13] They identify policy actors—state and local welfare officials—as influential in clients' access to postsecondary education. They also concede that the approaches of colleges and college systems are influential in how well they reach out to low-income populations. Thus, they show the ways in which access is limited for a population of nontraditional students. In contrast to these policy hindrances, Rebecca London argues that college attendance for this population is associated with improved outcomes, including those for employment and economic gains. Finally, Shaw, Goldrick-Rab, Mazzeo, and Jacobs demonstrate that the "work first" ideology impedes the progress of welfare recipients to education and training.[14]

Another equally disadvantaged population, the working poor, or "low-wage workers,"[15] are stymied in their efforts to improve their employment and earnings potential, largely because of problems of retention. Matus-Grossman and Gooden identify institutional barriers, such as lack of child care on site, lack of financial and employment incentives, and lack of advisement, as well as state welfare and workforce development programs, as contributing to access and retention problems. Several scholars see these populations under attack for reasons that stem from a) the low esteem and influence these students and their programs have within the institution,[16] from b) national and state reform proposals that threaten access, such as higher academic standards,[17] and from c) entrepreneurial pressures upon the community college that lead to an emphasis upon outcomes that result in diminished access for lower academic performing groups, such as women with children and the disabled.[18]

Institutional strategies are or are required to be variable in their response to student populations and the conditions of these populations. With limited resources—both fiscal and human—community colleges are hard-pressed to develop strategies and plan actions that will suit all specific populations. Programs, centers, initiatives, and support structures such as first-generation programs, minority student centers, case management, disabled student services, English as a second language integrative programs, Adult High School, and tuition and fee waivers are among the various strategies that are aimed at a broad spectrum of students. Most of these programs rely upon temporary funding and are both understaffed and underfunded, given the potential number of students who could participate but do not because the institution lacks resources.

Institutional Behaviors and Autonomous Agents

Institutional behaviors are a conglomerate of individual actions directed at students, and they are not necessarily consistent with strategy—that is, behaviors do not follow formal policy. They can be carried out by college faculty and administrators with personal agendas—that is, for motives that reflect the individual actor's values. These behaviors can have positive effects upon and outcomes for students. Some of these behaviors are focused specifically on the most disadvantaged students.

At Edmonds Community College in Washington State, the program director responsible for the Work First program (for welfare recipients) positions herself as a "gateway" keeper for the economically disadvantaged.

> I have now worked for the college almost five years and I was hired to work with the Washington State Work First program, and that program is designed to help TANF [federal welfare program] welfare clients, as well as low-income working parents to gain more job skills quickly, and then help them get a better job, better pay. . . . I am currently the director of the Work First program here on campus, so I am overseeing all of those programs with that same design. . . . Well, all the students that we are serving in our program are all parents. Sometimes we serve folks that you would call non-custodial parents, so they are paying child support, but they are still showing that they are parents. You've got some folks that have been on welfare for a while; some folks are working very, very hard at low-paying jobs and

they just need to get some additional training so they can do a little bit better. We have some folks who . . . have diagnosed and undiagnosed learning
disabilities that may be the reason they never did very well in school. In
fact, the vast majority of these folks did not do particularly well in school.
Some of them have to get their GED; some of them struggle to do your
basic learning. . . . We also serve limited English proficiency. . . . We are
now working with a vocational ESL program in health care and actually it
is going very well. They [health care faculty and administrators] are still
somewhat convinced that there are more on the "can't" side. These students
"can't do this," "can't do that." They're still leaning more toward that side,
where we are saying, "Oh yes they can. You just need to set the expectations
up there." They also want to teach to the higher ESL level, but we are saying, "What about these folks are coming up right behind them, aren't they
just as worthy?" So, we've divided it out, so that we are doing some pieces
with lower level and trying to get them to the next level. So it is a stair-step
thing. (Work First program director, Edmonds Community College)

The treatment of students for this program director is both at the individual
level—motivation and skill development—and at the group level—for a population with lower abilities than the norm. She sees her program as directed to
serving not just the welfare mothers but also their children.

It's all about the kids, all about the kids and making a livable wage for yourself and your family. I have to tell you that when these folks get through
these programs and they go out and get job, they come back to other graduations, we have little mini graduations, they come back to our graduations
and they talk about it. What a huge difference it has made in their lives.
How proud their kids are of them, because they are no longer humiliated
with food stamps at the checkout counter at grocery stores. They can actually have a decent apartment; they can actually buy a car. Wow. It has just
made a huge difference in their lives. . . . If you get this African American
population, they are unemployed; they can't pay their child support; they
can't access these programs. . . . They can't. . . . They are left out. . . . Now
the women with kids, they can use these programs and do. It was because
of the welfare reform act. The previous incarnation, JTPA [federal job
training program], could serve anyone . . . but that changed.

Her descriptions of both her work and goals suggest that from this perspective,
Edmonds Community College is a human services organization[19] attending to
the needs of its service population, one that is particularly stressed. Yet, because
of federal policy and regulations, her program cannot respond to those without
children: Males are excluded, as are females without children, unless they are
legal custodians of children.

She is not content with the social *status quo*, preferring to change lives by
presenting opportunities.

A couple years ago, I was involved with the School of Social Work at the University of Washington, but you know, I didn't want to do that. I wanted something that had more substance to it . . . it appealed to me to be working vocational. So I went to Seattle University in Voc[ational] Rehab[ilitation]. That seemed to have more of the cause and effect. You were actually helping people better themselves; you're not just kind of running in the same circles, running around, around, and around. You are actually helping those people focus their lives and get on a path; also, to challenge themselves. They can do better than they ever imagined. That's why I am here. Because they do. They are scared; they don't know what they can do. Those first few weeks are always very hard on them. . . . We always tell them, especially in our customized job skills training, "Once you get through this program, graduate, and go back to work, you will find that work isn't as hard." So, that is pretty amazing, and it continues to show me, and you have no idea what goes on sometimes in these people's lives while they are trying to make a perfect attendance record and stay in school, and all this stuff. We've had the absent parent kidnap the kids, domestic violence. You have emergency room visits because of severe health problems. It is amazing and the vast, vast, vast majority of them graduate.

This is indeed a rescue mission. She describes her work not as saving souls, but as missionary work nonetheless. She is not operating alone in her rescue mission, but is part of a group of like-minded faculty at her college.

> We have a very committed group of folks. Most of them have been in the business at one point or another, but everybody is pretty enthusiastic and it is great. We have a very team-oriented approach and that helps a lot.

Another example of individual behaviors directed at students with the personal motivation of the actor in play comes from a rural community college, an administrator who reflects upon her experiences as a counselor. In her work, she acknowledges the personal connection to students, as if personal and professional roles are indistinguishable.

> Having been the counselor, of course I was more closely aligned with some of those things related to retention. And two things, and this is totally opinion; I don't have anything to back it up. The fact that a counselor is assigned to an academic division, I think, makes for a very strong connection between the counselor and the student in particular program. "Those are my students." They know I'm their counselor. I see them the first time they walk in the door; talk about career exploration; and make a decision about which program to go into. I track them all the way through. They do end up having faculty advisors, but any time they're not sure they got the right advice from the faculty advisor, they come to their counselor. . . . But we were the ones that would look at the student, the time before they anticipate

graduating to do their clearance sheet, to make sure that they had met all the requirements and keep them on track to graduate. . . . The other thing that we do differently than some community colleges is how we teach developmental math. When I worked at North Fairfax, the thing that I worked on was looking at retention of at risk students. (Acting Director, Center for Business and Industry, Virginia Highlands)

Colleges operate in the context of state and federal policy, and sometimes institutional behaviors are constrained by these laws. One example has been the initiative of particular states to allow undocumented immigrants access to higher education at in-state rates, in the absence of any federal directive on the issue. As of 2005, nine states had enacted legislation that would allow such access, including Texas, New York, California, Washington, Illinois, Oklahoma, New Mexico, Utah, and Kansas.[20] Texas was the first state to cross the line, and it has been a pioneer in its approach to undocumented high school students who receive a diploma and have lived in the state for at least three years.[21] In 2001, state legislators enacted a bill to admit such students to public colleges and universities, allowing them to pay in-state tuition and qualify for state financial aid. Taking the opposite stance, Arizona citizens preventively approved an initiative called "Protect Arizona Now," which requires proof of U.S. citizenship in order to receive state and local services. The 2004 act also made it a crime for state and local officials to look the other way and not report a person seeking services without proof of citizenship. A middle-ground response is provided by North Carolina, which allows educational institutions and local boards to decide whether undocumented students from any jurisdiction should be admitted or not. Initially, admission for undocumented students was for noncredit programs only (such as ESL), but admission was expanded in 2004 to include credit-bearing programs at the out-of-state tuition rate. Seventeen more states have attempted to introduce legislation relating to this issue but have not been successful in passing it. Challenges and lawsuits from advocates on both sides of the issue are common. In the case of Arizona, the act was struck down by the courts and, until 2007, community college officials ignored the initiative. However, a similar initiative passed in late 2006 again penalizes undocumented immigrants in that state, nullifying public services for this population, including subsidies for community college education. In such an environment, nationally, some colleges are quietly making their own decisions regarding their actions toward their undocumented immigrant student populations.

Indeed, individual behaviors of institutional members often cross the line of institutional legitimacy or policy. In California, at Bakersfield College, a program administrator serves as a gateway for underserved populations as he works the system for students.

That [program] is an interservice agreement between the college and the state department of rehabilitation. The job of that . . . program is to place students with disabilities in jobs; in other words, take them through their job search to assist them and that is a very, very difficult thing to do. Being a minority, sometimes you think, I felt that it was difficult because being a minority, if you have a disability and you are really, really facing some challenges. So I took over that program and put those two programs together, again going at the same time they had a private industry council; they were doing what they called these summer youth employment training program, where they offered grants to higher education to work with individuals from age 14 to 21 over the summer. So I took over that program and added it to my evaluation, but what I did was change it. I limited the students to age 17, graduating seniors, because I also made it mandatory for them to do an internship. So, I turned it into a job, into an internship, and paid them to go to school, bought their books and everything. . . . Then we moved from early childhood education literacy program, combining with that a fostered youth mentoring program, then a tutorial assistance grant, and now we have a Smart team out on disabled. I went from just one program to almost four, and I wanted to do more, but they are not going to let me do it.

This administrator, an African American working in a predominantly white and Hispanic institution, altered the formal structure of college programs in order to fit the perceived needs of minority students, most of whom were Hispanic/Latino. In educating these students he was, in his words, performing community social service.

[I]n my study . . . what I found out is that we could go to the schools and find that most of the young people are in those schools, kindergarten through the eighth grade, they were reading, some of them were at the 35% percentile or below it, and as you know the goal was to have everyone reading at the fiftieth percentile by the time they completed the third grade. We were in trouble as a community, because if they could not get that fiftieth percentile by the time they completed the third grade the likelihood of ever reaching that point in high school was next to none. Alright, so what happened is, when they turn them loose and they want to go to the university, they are not eligible; they can't get there. You guys don't want them, because they are not eligible; they don't meet your criteria. So, then they come to us and we put them up in our skills lab. When they go up in those skills lab, if they are not successful very quickly, OK, then they lose interest and they are back out on the streets. Then you and I both pay for it, because they are doing the wrong thing. Sorry to get on my soap-box. (Program administrator, Bakersfield College)

With an almost insatiable appetite for establishing and administering programs in order to eradicate educational impoverishment, this college administrator operates his own educational system even though his main responsibility is cooperative education. The extent to which college and college system officials operate individually, sometimes beyond formal policy, is nearly endemic within the thirteen institutions addressed in my research. Such behaviors are not captured in other reports of research.

From the perspective of students themselves, the behaviors of college members serve as the bedrock of student performance and persistence, as noted by Liz at Edmonds Community College.

> Claire [writing instructor] worked with me. I was kind of lucky there because she was one of the instructors that didn't stay with the students that were ahead. She stayed with the ones that were behind, pushing them to the front. . . . She kind of became my mentor that winter: this was going to be my first real classes. . . . Because she goes, "You know what, you have your GED, don't stop. Why don't you work on your high school diploma?" . . . I had applied for financial aid. I had work study. . . . And they said . . . "Here are some places that are looking for some help. We're going to send you for interviews." I was like, "I've never been on an interview in my life." [The college administrator who hired me] decided to give me that break, and from that break, that just kind of knocked down a whole new wall for me. I started working on my high school completion and then I think I kind of ran through that little maze that everybody does—"Where are you going to earn the most money?" So I start going, taking computer classes and "if I have to do this the rest of my life, I think I'd rather go back to bar tending." . . . I decided to go into the office management. . . . From there I just started taking my classes. I finished my GED here; I finished my high school completion for my diploma here; I have one more class for my [associate degree in office management]. . . . I started adding on emergency responder because I ended up taking a permanent job here as a . . . security officer.

Among the many behaviors of college officials and system and state policy actors that affect nontraditional community college students, those that constitute individual action and are not necessarily role or rule bound are of considerable salience to nontraditional students, especially highly nontraditional students. These independent actors, whose individual agendas are often in play, I refer to as autonomous agents, using Lipsky's concept of "street-level bureaucrats"[22] as a guide. Autonomous agents at various levels, from the program level of faculty and coordinators to the level of state system presidents, enact personal preferences and agendas either directly or through policy. Additionally, organizational power arrangements or configurations[23] privilege some organizational members and influencers so that they have substantial power in the decisions and actions of organizations. Such influence may be vested within groups or individuals. Take as one prominent example my conversation with the president of the North Carolina Community College system ("A" constitutes the president's responses and "Q" my questions):

A: But today at, I think, one-thirty, we will have the first meeting of the advisory committee of the Latino Initiative. I went to the Steve Smith Reynolds Foundation and got two-year funding to do a Latino Initiative, with the stated purpose of taking Hispanic students from ESL into skills development, and I don't have the figure, but it is a very high percentage, better than 90%, in my opinion, of the Hispanic enrollment on community college campuses is in ESL, and they simply do not take advantage of skill development after they get ESL. What I wanted to do was to develop strategies for making community colleges more Latino-friendly: developing programs that would accommodate their special needs and move them into special development, both in continuing education and curriculum.

Q: So this advisory committee is going to be for the community colleges?

A: A part of this grant will be a statewide initiative of the systemwide office, and we will be hopefully developing programs and strategies that campuses can use. Again I can't mandate it, but what we are hoping is that what we will come up with will be filling a need for colleges. I think the colleges want to be more than just ESL factories; they just don't know how. It is a cultural thing, and we believe that this advisory committee is heavily Hispanic and we want them to sensitize us, to spell out needs that are unmet, the whole works. And then, from their suggestions, we will be developing the strategies and programs.

Q: Where are you going to get the money for these programs or initiatives?

A: The initial initiative is a grant from the Steve Smith Reynolds Foundation.

Q: And you procured that?

A: Yes, I got that. We have one and a half positions, a director and a part-time support person and this advisory committee, and we will develop programs, strategies that may lead to our going to the general assembly for discrete funding. It may lead to further grant applications; it may lead to colleges simply doing things differently with the money they have in order to be more responsive to the Latino needs. My guess is that most of what we recommend will not require additional funding. It will simply be sensitizing campuses and their staffs, helping them see ways to better serve, attract, retain the Latino student.

Q: And, besides you being a good humanitarian, was there great pressure from Latino groups?

A: No, there was not; there was zero. This was simply my . . . it was really a big personal initiative, because I did not feel that we were meeting the needs. When I looked at our enrollment data and saw 90-plus percent of our Hispanic enrollment was in ESL, what I saw in the census data was a 400% growth in the Hispanic population, then I thought how small our enrollment was even with ESL and I knew we were not meeting the need. So I went to the Steve Smith Reynolds Center. It took me two cycles and then they gave me less than half of what I asked for and so we—you know how it is.

Q: Now, if I am wrong, correct me, but how do you then get the student from basic skills, which is noncredit, noncurriculum, into curriculum?

A: That is the purpose of this grant. That is what I want to know and I don't know. (President, North Carolina Community College System, January 21, 2004)

It is patently evident that this official has taken the authority and leverage of his position and used that to fulfill a personal agenda of improving higher education opportunities for Latino students, many of whom are not legal immigrants and find themselves stymied by institutional regulations and federal law.

In another example, the chancellor of the nation's largest community college system—California—expresses not only a personal agenda for the state's community colleges but also awareness of his influential position. During my meeting with the chancellor, he asked if he could stand and walk around his Sacramento office while he talked with me, and this he did for almost two hours.

I guess I have a personal agenda. . . . I want to see the colleges much more aligned with business leadership in the state. . . . [J]ust this very day it was announced that I was appointed to the state board, to the board of state chambers of commerce. I am the first chancellor that has ever been so appointed and I sought that out. I know people that obviously got me that appointment, but I will now be on the state chamber along with President Dynes of the University of California. So that will, in time, give me a chance to put a personal agenda in and realign this thing and try to get it more embraced by business. I am forming a chancellor's cabinet, which will be senior corporate managers from around the state. They will cough up some money to be a part of it, but I will go have lunch and talk nice to them. It will be like a kitchen cabinet: We will close the door and plot agendas and things that might help the business community. At the same time I will be after them for raising money. We have never raised much scholarship money here at all, in my opinion. We should have hundreds or millions endowed; we don't have anything. So, there is a whole thing . . . that is a personal agenda. . . .

It may just help our kids get better jobs. The other part I have is just at the opposite end of the spectrum. I have a great concern. I am a native Californian. I have all the good and bad that goes with that, but one of the things that troubles me greatly is the welfare and the receptiveness of our, California's, society to mainly migrants, low-income people, but also to people who are not migrants, people who are simply typically lower SES kids, intercity kids who never find the first rung on the social ladder. I don't know if you know this number or not, but today in the 2000 census we had over a million young people between eighteen and twenty-four who did not possess a high school diploma. Now, today that is probably 1.2 million. I do also know that in California the unemployment rate—they call it the youth unemployment rate, that is, the eighteen to twenty-four [year old] unemployment rate—is about 22%. So we have an enormous [problem]; it is like a big dirty secret . . . [the] number of people, I mean, 1.2 million people, and that is almost as many students as we have in our whole state. And they are out there without high school diplomas. And . . . some of them are washing cars; a lot of them are just wandering the streets . . . I mean they are out there and they are clogging the health care clinics when they get hurt. They go into the emergency rooms; they are on any form of public assistance that they can find. I mean they are not paying their own way. They are not paying taxes; they are not helping the school systems; yet they are having kids. I mean . . . it is a huge problem and I don't see any better doorway to the first rung on that social ladder than the community college. Frankly, today we are not structured, nor are we funded to address that population. We don't really address them. Some of them wander in; we assessment test them; if they test below the tenth grade they have to go to noncredit/remedial stuff. . . . We know how, I mean there is research on how to help these people overcome their skill deficiencies, but it is expensive. It takes small classes; it takes six hour long classes; it takes daily classes, all kinds of attention to tutorials. . . . You can't do it for hardly the full . . . portion, let alone half. So we just don't do it. We do give these classes, but we pack eighty people into a class, give them a tape and a headset and say sit there and listen to the stuff and tell me how it was. I mean, it is miserable. . . . The whole bottom end of that system is broken, and that is a personal agenda of mine, to do something about that, and we have already started. . . . We are going to be putting a group together to study this population; study the known solutions; study the structure it would take for us to do it; and then I will go to the legislature. That is how I influence policy. I can take a budget change proposal over to the legislature and I can actually, through this office, get a funding stream to start pilot programs. That has been done quite a bit. We probably have right today, I'd say somewhere between ten and twenty million [dollars] in pilot programs flowing through this office, out to districts that are doing. . . . So, we have the power and the ability to do that. We spin things up. Now, once again, the riddle of that is, like it is with any grant-funded activity, "How do you pick the winners and then plant them, so they don't just quit at the end?" That remains to be seen, but at least we have the ability to do those things. (Chancellor, California Community Colleges, 2004)

Both critic and champion, the chancellor of California's 108 community colleges in 2004 articulates the failings and shortcomings of state policies and funding practices. The lower socioeconomic strata of the state, comprised of eighteen-to twenty-four-year-olds, is sapping the economic life out of the state. His intention is to resuscitate "the whole bottom end of that system," and it is his "personal agenda . . . to do something about that, and we have already started."

Employing what Mintzberg terms the "missionary configuration" of power,[24] that is, aligning external and internal influencers through strongly held belief systems, the two system executives are able to exercise power for specific issues and causes. They enable a large immigrant population in the case of North Carolina and California to move beyond their starting points of educational deficiencies to higher educational and, ultimately, economic levels. Compared with the influence of individual teachers or administrators at community colleges, these policy executives can affect large populations through their actions.

Institutional Effects

While some scholarship suggests that institutional effects are far outweighed by student effects,[25] this may not be the case for highly nontraditional students who are severely disadvantaged by their backgrounds. Large-scale interventions brought about by government legislation—including state propositions to curtail services to undocumented immigrants—can have profound effects upon large numbers of students. Arizona is a recent case where public services to undocumented immigrants were viewed as illegal acts. Similarly, in North Carolina, at the time of the initial portion of my investigation, undocumented immigrants could not register for credit programs at community colleges or universities. (This law has now altered.) So, too, can institutional actions have important consequences. Institutional effects are or can be significant for those actions—tacitly agreed to by institutional members—that ignore these system and state policies and laws that exclude or disadvantage populations from equal participation in postsecondary education.

At the several colleges I observed, college employees—from presidents to mid-level administrators, to faculty and staff—relied upon personal ethical standards and ignored state or federal policy so that students were not harmed. They resorted to "illegitimate" behaviors or "spin" rationales for their actions to accommodate students, to ensure that students could enroll in credit programs even though they were undocumented immigrants, to ensure that they could finance their education even though they might not qualify for grants or scholarships, and to ensure that programs and instructional attention were available even in the face of legislative and public political opposition or indifference. These efforts effect change both for individual students and student groups: They frustrate, curtail, impede or indeed nullify the expected trajectory of student participation and achievement based upon student characteristics.

In that the behaviors of autonomous agents are not necessarily legitimate or institutionally sanctioned, or within existing policy norms, these behaviors can be idiosyncratic. That is, they are not systematic or formally institutionalized, and they do not apply to all groups or populations. Indeed, these behaviors at the institutional level—at individual colleges—are aimed largely at individuals not groups. Thus, 10 percent of an eligible population may gain advantage— through a tuition waiver, for example—but the other 90 percent are not as fortunate. While autonomous agents play a significant role in how colleges affect students, their behaviors and the resultant actions can be deleterious as well as advantageous. They can pursue their own agenda and ignore other equally important issues. Or their behaviors can lead to unintended consequences, such as false hopes to student who begin to think that they can advance socially and economically, when in fact they may be denied access to employment (e.g., because they do not have legal immigrant status) or to further education (e.g., because they do not have fiscal resources). Nonetheless, in a neoliberal environment, autonomous agents who favor fairness and advantaging the disadvantaged become the rescuers of large numbers of students from dismal prospects.

Conclusion

Intentions do not necessarily yield results or the expected ends. Institutional strategies, on the one hand, are plans, and they may be enacted in ways unanticipated by the designers. The outcomes of organizational strategies are sometimes elusive, sometimes fictional, and sometimes deleterious.[26] Strategies of community colleges can be planned responses to global forces, including economic markets, immigration patterns, new communication technologies, and federal and state policies.[27] Because the community college is a mission-driven institution, with open access to educational opportunities as one if its principles,[28] it sits uncomfortably with neoliberal policies that favor economic capitalist ends at the expense of community cohesion and the welfare state. Indeed, institutional members as well as community college system bureaucrats oppose curtailment of the human and social services that community colleges offer to students and community members. Because institutional strategies are formal, rational, bureaucratic responses, nested within a legitimated system, they are not necessarily useful for counteracting the perceived ill-effects of neoliberal policies or any policies that take resources, human or fiscal, away from the community college. Thus, autonomous action, not strategy or formal institutional actions, becomes the method whereby mission is pursued. In this vein, justice as fairness[29] is not accomplished as a systematic action of the community college but rather by individual action or the informal collectivity of individuals. These autonomous actions are idiosyncratic and their results, while beneficial to some, are not universally applicable.

Notes

1. Thomas R. Bailey and Vanessa Smith Morest, "The Organizational Efficiency of Multiple Missions for Community Colleges" (New York: Teachers College, Columbia University, 2004); "California Community Colleges System Strategic Plan. Education and the Economy: Shaping California's Future Today" (California Community Colleges System Strategic Plan Steering Committee, 2006); W. Norton Grubb and Marvin Lazerson, *The Education Gospel* (Cambridge, MA: Harvard University Press, 2004).
2. Arthur Cohen and Florence Brawer, *The American Community College* (San Francisco: Jossey-Bass, 2003); John Frye, "Educational Paradigms in the Professional Literature of the Community College," in *Higher Education: Handbook*

of Theory and Research, ed. John Smart, 181–224 (New York: Agathon, 1994); David F. Labaree, "From Comprehensive High School to Community College: Politics, Markets, and the Evolution of Educational Opportunity," Research in Sociology of Education and Socialization 9 (1990): 203–40.

3. Frye, "Educational Paradigms."

4. Burton Clark, "The 'Cooling-out' Function in Higher Education," American Journal of Sociology 65, no. 6 (1960): 569–76; Kevin Dougherty, The Contradictory College (Albany: State University of New York Press, 1994).

5. Clifford Adelman, Moving into Town—And Moving On: The Community College in the Lives of Traditional Age Students (Washington, DC: U.S. Department of Education, 2005); Grubb and Lazerson, The Education Gospel; Tania Levey, "Reexamining Community College Effects: New Techniques, New Outcomes" (paper presented at the Association for the Study of Higher Education, Philadelphia, PA, November 2005).

6. Cohen and Brawer, The American Community College.

7. John Rawls, Political Liberalism (New York: Columbia University Press, 1993).

8. Alan Seidman, ed., College Student Retention: Formula for Student Success (Westport, CT: Greenwood, 2005); Vincent Tinto, Anne Goodsell Love, and Pat Russo, Building Learning Communities for New College Students (State College, PA: The National Center on Postsecondary Teaching, Learning, and Assessment, 1994).

9. John E. Roueche, Eileen E. Ely, and Suanne D. Roueche, In Pursuit of Excellence: The Community College of Denver (Washington, DC: Community College Press, 2001); Roueche and Roueche, High Stakes, High Performance: Making Remedial Education Work (Washington, DC: Community College Press, 1999).

10. Katalin Szelényi and June C. Chang, "Educating Immigrants: The Community College Role," Community College Review 30, no. 2 (2002): 55–73.

11. Wendy Schwartz, Immigrants and Their Educational Attainment: Some Facts and Findings (Washington, DC: ERIC Digest 116 [ED 402398], Office of Educational Research and Improvement, 1996).

12. Jerry A. Jacobs and Sarah Winslow, "Welfare Reform and Enrollment in Postsecondary Education," The ANNALS of the American Academy of Political and Social Sciences (March 2003): 194–217; Rebecca A. London, "The Role of Postsecondary Education in Welfare Recipients' Paths to Self-Sufficiency" (Santa Cruz: University of California–Santa Cruz, 2004).

13. Christopher Mazzeo, Sara Rab, and Susan Eachus, "Work-First or Work-Study: Welfare Reform, State Policy, and Access to Postsecondary Education," The ANNALS of the American Academy of Political and Social Sciences (March 2003): 144–71.

14. Kathleen Shaw et al., "Putting Poor People to Work: How the Work-First Ideology Eroded College Access for the Poor" (unpublished manuscript, 2005).

15. Lisa Matus-Grossman and Susan Tinsley Gooden, "Opening Doors to Earning Credentials: Impressions of Community College Access and Retention from Low-Wage Workers" (paper presented at the annual research conference of the Association for Public Policy Analysis and Management, Washington, DC, November 2001).

16. Grubb, Noreen Badway, and Denise Bell, "Community Colleges and the Equity Agenda: The Potential of Noncredit Education," The ANNALS of the American Academy of Political and Social Sciences (March 2003): 218–40.

17. Penelope E. Herideen, *Policy, Pedagogy and Social Inequality: Community College Student Realities in Postindustrial America* (Westport, CT: Bergin and Garvey, 1998).
18. Kathleen Shaw and Sara Rab, "Market Rhetoric versus Reality in Policy and Practice: The Workforce Investment Act and Access to Community College Education and Training," *The ANNALS of the American Academy of Political and Social Sciences* (2003): 172–93.
19. Yeheskel Hasenfeld, *Human Service Organizations* (Englewood Cliffs, NJ: Prentice Hall, 1983).
20. American Association of State Colleges and Universities, "Policy Matters: Should Undocumented Immigrants Have Access to In-State Tuition?" AASCU newsletter 2, no. 6 (2005).
21. Radha R. Biswas, *Access to Community College for Undocumented Immigrants: A Guide for State Policymakers* (Boston, MA: Jobs for the Future, 2005).
22. Martin Lipsky, *Street-Level Bureaucracy* (New York: Russell Sage Foundation, 1980).
23. Henry Mintzberg, *Mintzberg on Management: Inside Our Strange World of Organizations* (New York: Free Press, 1989).
24. Mintzberg, *Power In and Around Organizations* (Englewood Cliffs, NJ: Prentice Hall, 1983).
25. Thomas R. Bailey et al., *The Effects of Institutional Factors on the Success of Community College Students* (New York: Community College Research Center, Teachers College, Columbia University, 2005).
26. Robert Birnbaum, *Management Fads in Higher Education: Where They Come From, What They Do, Why They Fail* (San Francisco: Jossey-Bass, 2000); Mintzberg, *Rise and Fall of Strategic Planning* (New York: Free Press, 1994).
27. John Levin, *Globalizing the Community College: Strategies for Change in the Twenty-First Century* (New York: Palgrave, 2001).
28. Quentin Bogart, "The Community College Mission," in *A Handbook on the Community College in America*, ed. George Baker, 60–73 (Westport, CT: Greenwood, 1994); Cohen and Brawer, *The American Community College*; Robert Rhoads and James Valadez, *Democracy, Multiculturalism, and the Community College* (New York: Garland, 1996); Roueche and George A. Baker III, *Access and Excellence* (Washington, DC: Community College Press, 1987).
29. John Rawls, *A Theory of Justice* (Cambridge, MA: Belknap Press of Harvard University Press, 1999).

Chapter 7

Continuing Education
and Lifelong Learning

Introduction

The topic of continuing education arises in relation to nontraditional students because of several associations: first, because of the ties of continuing education to adult learners; second, because of the connotations of continuing education to returning students; and third, because of the connections in both the scholarly and practitioner literature between continuing education and lifelong learning.[1] This chapter connects both continuing and community education programs and courses at community colleges to nontraditional students. Continuing education and community education have distinct histories in community colleges, yet over the past three decades these distinctions have eroded. For example, with state funding reductions for community education, courses offered as community education have become centers of cost recovery, not unlike courses offered to generate revenue. Self-help courses such as "Starting your own business" are no longer state supported, and student tuition for these noncredit courses, theoretically, pays for instructional costs. Additionally, meanings of the two terms vary from state to state; or, indeed, in some states, community education has lost its affiliation with the community college, finding a home in community centers or with private providers. Generally, community education has signified those courses and programs that are avocational, with a recreational slant, noncredit-bearing, and offered on the basis of community interest. Continuing education, in some distinction, has signified those courses and programs with a vocational or further educational purpose, which may or may not be credit-bearing. Some colleges use the continuing education unit of the institution to offer certificate programs that lead to an externally awarded credential. Examples of programs and courses that are typically referred to as continuing education include contract training for business and industry, Basic Skills, Adult High School, GED, English as a second language (ESL), Special Education (for disabled populations), and a host of vocational certificate programs from truck driving to medical office receptionist. Community education typically includes courses and programs such as Elder Hostel (or other comparable programs for seniors), recreational activities (e.g., tennis, yoga, cooking), educational travel, and self-help guidance (e.g., income tax, first aid, and weight loss).

The focus of this chapter is primarily upon the continuing education unit within community colleges and its impact upon students. The chapter shows how continuing education programs structure nontraditional student learning. The discussion examines the structures and purposes of continuing education and considers how these factors affect not only access to the institution but also outcomes for nontraditional students. Do outcomes constitute justice, particularly for disadvantaged students, as a result of continuing education? This question is addressed through the scholarly literature's critique of the concept of "lifelong learning" and the discourse that surrounds this concept. Economics is one starting point for this discourse.

Unlike other sectors of higher education, such as research universities, community colleges have two main sources for revenue—the public, in the form of the state or the local community, or students themselves, through tuition and fees. The state has by far been the major provider of finances for community colleges since the mid-1970s. Most state systems use some type of formula funding approach where student enrollments are the critical variable for rises and falls in state appropriations. However, nationally, the state's proportion of funding the entire operation of the community college has diminished over the past twenty years, from 60 percent of the total in 1980 to 44 percent in 1997. This has been accompanied by a rise in local funding, which has increased from 13 percent to 19 percent of the total, and tuition and fees rising from 15 percent to 21 percent of the total.[2] With limits upon taxation and the decreasing fiscal role of the state for funding, community colleges have emphasized strategies and actions that will generate revenue. Students are the targets of these new strategies.[3]

This is a departure from the community college's traditional pattern of behavior, which, although focused upon student enrollments, was motivated by the institution's mission of expanding educational and training opportunities for the community.[4] As a result of this shift, organizational behaviors at community colleges imitate behaviors found in the business sector, with economic goals dominating institutional strategies and actions.[5]

The economic imperative for community colleges, combined with the mission of access, may indeed imperil accommodation of students in the sense that accommodation may not constitute justice but instead the customary "open access."[6] As will be discussed below, continuing education is increasingly used by colleges to solve both the problem of insufficient revenues and the limitations of access within traditional educational structures.

Continuing Education

It is estimated that over 50 percent of adults (those over twenty-four years of age) who undertake some form of postsecondary education do so outside of the traditional credentialing framework. If credentials are involved for this population, it is an industry or company credential, not a postsecondary education institution's credential.[7] Nonetheless, large numbers of these students fulfill credential requirements at community colleges. This figure is significant on two counts: First, this group is not captured in national educational data sets (by National Center for Education Statistics, or NCES) to reflect student behaviors in postsecondary education; and second, our understanding of postsecondary education students is severely limited if we ignore those in continuing education programs where degrees, for example, are not the "coin of the realm." Increasingly over the past three decades, continuing education has had a decidedly labor market focus, and employers are important players in this activity as they subsidize their employees in training programs offered through postsecondary education providers.

Structures and Purposes of Continuing Education:
Student Access and Revenues

Richard Bagnall argues that there have been three major sentiments that have shaped continuing education: 1) "commitments to individual growth and development," 2) "commitment to social justice, equity, and social development," and 3) "commitment to cultural change."[8] These sentiments are appropriate to the advancement of nontraditional students, particularly to the advantaging of the disadvantaged. Bagnall, however, concludes that these sentiments have given way to economic determinism with students viewed as commodities. Patricia Gouthro and others[9] concur with this conclusion: Students in higher education are viewed as a consumable. The consistent and traditional view is that adult learning and continuing education have not been responsive, historically, to economic markets or subservient to consumerism. Instead, the values of adult learning have featured self-expression, identity development, and social criticism, as well as skills that improve individual positioning in the workplace.[10] The purposes and structures of continuing education for adult learners have, over the past two decades, become consumer oriented, taking such forms as contract education and training and distance education or distributed learning, in a response to the "risk society," defined by Ulrich Beck in 1992 as a society where individual responsibility and individual, not social, needs are ascendant.[11]

This is in contrast to the historical pattern of community colleges that suggests these institutions have served low-income students through their adult education or continuing education units. Largely through noncredit education funded by local governments or the state, community colleges brought affordable education and social development to local communities, many of which are in rural areas where there is no other postsecondary educational institution. But with the loss of state funding, pressures from employers for a trained workforce, and shifts in government—both state and federal—policy, noncredit programs gave way in both number and importance to credit programs and those courses and programs that could be credentialed externally.[12] With this scenario, it is not difficult to see that a class of noncredit students—those engaging in courses and programs without goals for credentials or specific workplace training—become invisible or segregated from institutional life. This is particularly the case for developmental, English as a second language, and Adult Basic Education students. Thus, the programs are vulnerable to elimination by resource providers, as noted by a dean at Pima College in Tucson, Arizona.

> [T]he reason that I think it is vulnerable is because the people that it serves, or it is designed to serve, are the weakest political constituency in the United States by virtue of any indicator you can use, and that is why it has been a marginalized program for years, serving the marginalized. . . . [T]hat is why it is the most vulnerable program right now. (Dean of Adult Education, Pima College, Arizona)

Noncredit, then, flourishes when it is tied to training for business and industry—it can yield profits; it is more flexible than credit programs; and it does not have the same reporting requirements as credit programs. The chancellor of the Maricopa County Community College District in Arizona notes that his institutions are moving to increase their noncredit programming because it is outside state polices and can achieve greater revenues than credit offerings.

> For example, Motorola: We will do contract training [noncredit courses]; we will have a contract rate; and we can do some classes that are fully paid for. They are profitable, but there are other courses that we do for credit, that we won't make any money on because we will only charge them the tuition and fees. . . . If Motorola sends up a class of 120 people, sixty of those people are here within the state. They actually live in the state. The other sixty are from all around the country, [and] we will charge them a differential rate. We will charge the one the out-of-state rate; the others the in-state rate. But we won't charge them much more than the premium. . . . We probably won't make a big profit out of it . . . if it is totally a noncredit class. If it is totally a noncredit class we will charge whatever we think the markets are. . . . A lot of people want retraining now and they really don't need credit. . . . Most of our classes are designed for credit. So that is causing us a problem. So we are beginning to look at other types of modules, other types of setups, that would allow us [to pick up business]. . . . I would say continuing education . . . will be our bread and butter in terms of how flexible we are to be able to respond to the market, and this is where state policy can come into effect as well. What we are asking the governor to do is to look into ways that any legislative mandates or compliance issues around what community colleges can and cannot do, in terms of partnerships . . . [and] that she assist us in terms of changing the law, if necessary, to meet these needs. It could be reporting requirements; it could be issues of competition that they could control; it could be taxing issues. How can we invite companies to work with us [so] that they

would get a benefit from the state? . . . We will have a tendency to move to noncredit because we have not been funded for credit . . . over the last three years. So, it doesn't really help us to try and do credit. As a matter of fact, it is a push in most cases. It takes a lot of time and effort. If we just build it for noncredit and staff it for noncredit, and just exchange, it works out better. (Chancellor, Maricopa County Community Colleges)

Preoccupied with financing institutional operations as state funding shrinks and demand for services expands (especially evident in the progressive growth of enrollments and demographic alterations),[13] college leaders resort to practices consistent with neoliberal ideology. This ideology pays homage to the economic marketplace and promotes the reinforcement of competitive structures for economic behaviors.[14] Furthermore, as the chancellor of Maricopa County indicates, government is a facilitator of economic market freedom, not a vehicle for the economic security of individuals.

The dean of community advancement at El Camino Community College in California identified the separation between those units within the college that are revenue generating and those that are not: "The more we can become separate the better off we are." The reasons are clearly economic.

It is because we have to operate like a business. We have to be entrepreneurial so we can be ready if Boeing comes up and wants training. They, like every other business, are going to want to start up next week. They want the best people to do it; they want it to cover a midnight to 7 a.m. shift. . . . They want it to be short term, like eight weeks. So we have to hire people to do it, you can't hire faculty [from the credit units of the college].

The separation permits not only fewer bureaucratic constraints but also production unfettered by academic oversight: no faculty committees to question curriculum or instruction; no union contract to regulate working conditions; and no disciplinary body or representatives to ensure compliance with academic professional standards or practices. Students in such programs and courses are deprived of education and training that is demonstrably equal to that provided in credit instruction. Indeed, these students are excluded from education and training that is not part of the corporate ethos, as education and training are infused with skills that are needed by the business or corporation to remain competitive.

Structured Learning through Continuing Education: Contract Training

Learning is thus structured for economic purposes: for workforce development and for individual skills required for initial employment, retraining, or career advancement. Continuing education units in higher education institutions have become the structures through which these education and training ends are met, bringing some institutions profit and most institutions revenues in order to support their units. In the community college, contract training has become a mainstream activity to serve business and industry.[15]

The definition of "contract training" offered by the United States Government Accountability Office (GAO)—employee training provided under contract to businesses, government entities, or other employers, including for-credit and noncredit courses—is consistent with basic definitions of contract training offered by community college scholars.[16] Arthur Cohen and Florence Brawer use the term to refer to instruction that is provided for specific occupational purposes in three categories: 1) training designed specifically for the employees of certain companies; 2) training for public-agency employees; and 3) training for specific groups such as unemployed people or people on welfare. Training is provided for incumbent workers in targeted industries or for prospective workers in new and developing fields. The GAO study found that more than three-fourths of community colleges are involved in contract training programs designed to enhance the skills of incumbent workers. It concludes that offering existing or customized programs on a noncredit basis enables community colleges to use shorter training periods and allows them to adjust to local training needs. Courses are offered on both a credit or noncredit basis.

> Regardless of whether programs were credit or noncredit, schools most frequently offered occupational, professional, or technical training programs in three fields projected by the Department of Labor to have high growth in future years—health care, business, and information technology.[17]

Beyond flexibility and responsiveness as reasons for employers choosing community colleges for training of employees, the primary reason that businesses turn to community colleges for contract training, as identified by Jenkins and Boswell, is that contract training is subsidized by state government, and in many instances it is offered without charge to new and expanding businesses. State policy therefore becomes a key factor in determining the level of community college involvement in contract training. Business leaders, forced to address the increasing skill demands placed on incumbent workers by advancing technology and the shortage of trained workers in new and expanding industries, have played an instrumental role in initiating state policies that support contract training, while the community college systems have been willing accomplices in the lobbying effort. As a result, community colleges have been designated as the lead agency to provide workforce training in at least nineteen states, and they are the most prominent players in the remaining systems.[18] Dougherty and Bakia see this trend as both the result of pressures from business and the "evangelizing efforts of community college associations," while there is also government encouragement through "exhortation and financial incentives."[19] As community colleges require more resources to sustain their level of services, the revenue-generating units are viewed as critical to their mission.

> [B]asically we go out there and make money. We develop contracts for business and organizations. We provide services that range from the initial consulting services to direct training services usually deliver on-site

for the company. . . . I've been showing the college and they're looking at cutting things and I'm saying, "Well, don't cut us." We're your best possible chance to start making some serious money around here. . . . I'm very much entrepreneurial-minded. I want to make money, and the college . . . should be thinking that way. They're allowed to. They can make money. They can charge fees. . . . There's no prohibition from us making any money that's reasonable if, in my view, we're serving the mission that we're supposed to serve. (Director of Corporate and Community Services, Bakersfield College)

Despite philosophical struggles that exist over the use of public monies to support private enterprise, state legislators continue to fund customized contract training programs to attract new industry and to improve the skills of incumbent workers. Businesses benefit from the flexibility and responsiveness of community colleges, which commonly provide custom-tailored courses to fit their unique needs. The colleges benefit from increased enrollments, additional revenues, and strengthened ties with local business leaders, who may in turn lobby legislators on their behalf. Furthermore, in some states, such as North Carolina and South Carolina, where the community and technical colleges were organized historically with economic development as primary objectives, activities such as contract training are consistent with their original statutory mission.[20]

Contract training and workforce development are key activities of community colleges. That raises the question of the role of the community college in the new economy. The institution can be viewed as a vehicle of neoliberalism, appropriating the concept of lifelong learning and shaping the concept with a decidedly economic purpose. Indeed, we might conclude that continuing education for a subset of adult learners is substantially workforce development, couched in the euphemism of "lifelong learning."

Lifelong Learning

Lifelong learning has two primary connotations. First, the concept is associated with individual development over a lifespan and the accompanying need for knowledge to compliment that development and engage individuals with their society over time.[21] Second, the concept is associated with workforce training needs of the economy, whether local, regional, national, or international, and the need for learning is to fit the individual into the changing economy and workforce so that economies can maintain their competitiveness.[22] In this sense, the learner or student is an instrument, used to satisfy the economic needs of the private sector—to provide a globally competitive workforce as labor; to fill professional positions of an advanced economy; and to occupy a place at the business and techno-table where strategies for adapting to postindustrial production are developed.[23] This condition is the logical extension of an institution—the community college—that has aimed its purposes and curriculum to please employers.[24] "Lifelong learning" has been the clarion call of adult educators and community college officials as well as observers such as policymakers to promote both increased participation in college education and the new niche of stature for community colleges. The reasoning offered is that in a complex, postindustrial society, knowledge expands, and that in order to cope—personally, socially, and vocationally—with this new knowledge, all must continue their education and their learning throughout their lifespan.

From the perspective of economic production, however, lifelong learning has a much less personal focus and suggests that forms of production alter, and that in order for business and industry to remain competitive, their workforce must be "retooled" or retrained. Furthermore, with a rapid shift in jobs and employment, workers must continually retrain just to qualify for jobs. Indeed, lifelong learning is equated with a career, which is seen as central to individuals' identities.[25] Thus, individual identity and the meaning of personal and social experience are framed educationally within an economic context.[26] Even the highly vaunted 'learning communities' concept as a goal for educational institutions is equated with an adaptive approach for economic alignment.[27] Lifelong learning, then, from this vantage point, is a concept compatible with neoliberal ideology. Through this economic imperative, the concept of lifelong learning has entered the front door of higher education institutions, no longer on the periphery of operations and missions.

The ascendancy of lifelong learning in community colleges arises from and reinforces a "consumer-centric model" that places the institution in an economically competitive environment.[28] This environment has a decidedly business-like orientation where traditional understandings of the academy are misunderstood, ignored, or reformed. Instead, business solutions are proffered for problems that may be essential characteristics of colleges and universities, such as ambiguity, organizational slack, and diversity of views and values. It is not so much that politics dominates the academy as some have claimed,[29] but that economics prevails. Academic governance and institutional autonomy are threatened when resource generation and efficiency measures take a central role in universities.[30] In the community college, however, academic values and traditions are both weak and when present underdeveloped.[31] To some extent, this is understandable since community colleges have multiple purposes, many of which are not directed to the pursuit of knowledge or intellectual or cognitive development of students.[32] Job training is arguably a main feature of the institution, and within that context, academic freedom, for example, is not a precious commodity.

Economic behaviors of community colleges have become more prominent with the scarcity of government funds to finance ever-expanding programs and growth in student numbers. In the fall of 1990, there were 4.9 million students in public community colleges; in 1995, the figure was 5.2 million; and by fall 2001 the figure was just under 6 million.[33] These figures reflect a 24 percent growth in just over a decade. Credit enrollment has now climbed beyond the 6 million figure.[34] The pursuit of revenue sources led community colleges to such actions as contract training, international education, and cost-recovery programs.[35] While the outcomes of these actions do include revenue generation, they also include program, mission, and institutional modification.[36] The ever-increasing number of part-time faculty and the expansion of distributed learning are among those organizational changes that have altered institutional mission.[37] These changes suggest that economics might be undergirded by ideology.[38]

To what extent are community colleges fostering a neoliberal ideology through the rationales and practices of college officials who promote lifelong learning as service to business and industry? Do college administrators responsible for institutional policy and curriculum deliberately treat students as instruments of economic competition? I use the analytical frameworks of neoliberalism and Casey's concept of "corporate cultures,"[39] as this concept pertains to higher education institutions, in order to understand how community colleges frame and enact lifelong learning. Casey argues that the goal of corporations in educating and socializing employees is to remake workers so that they are "designer employees" who internalize the values and practices of the workplace and carry these values over into their personal life. This education and socialization aim includes the preparation of corporate workers who become compliant employees: "dependent, overagreeable, compulsive in dedication and diligence, passionate about the product and the company."[40] Specifically, I discuss the extent to which community colleges serve as proxies for corporations and businesses in their preparation of employees.

This discussion involves several interconnecting issues. On the one hand, there is the economic function of community colleges, both for city, state, and national economies and for individual economic well-being. On the other hand, the development of a workforce for the private sector—to serve the needs of private capital—may not necessarily coincide with public or individual citizens' interests. Furthermore, in training or preparing students for the workforce, community colleges make choices on the level and scope of training and whether training includes more than skill development, such as intellectual or cultural development. No doubt, there are pressures from both employers and government funders for short-term, job-specific training.

I examine the views of administrators at senior levels of these institutions and at the state level. These administrators have policy as well as practitioner roles: Their personal views on college goals and actions can be influential in the actual operations and outcomes of their institutions and their state's community colleges. One set of issues is articulated in interviews with four administrators in three states—one in Arizona, one in California, and two in North Carolina. Two of these administrators work within specific colleges as managers. The other two are statewide executives for community colleges. In the first, the dean of instruction offers her characterization of the goals and behaviors of community college students, which include a predominantly workforce orientation. In her explanation, she sees college education as instrumental both for the students and for her institution. The students pursue skills to obtain jobs and the institution prepares students for the workforce. While there is reference to university transfer, the assumption is that this education too will result in economic benefits through employment.

Community college students come here for very different reasons than to graduate. Most are not here to graduate from this institution. Most are here to gain the skills to go to work; most are here to gain enough credits to move on to the [university]; most are here because they are in a job that requires upgrading. So, in that regard, all three of those outcomes are things that will improve their economic futures, sometimes very immediately. I am not at all interested in what our graduation rate is. I am not at all interested in that. What I am interested in is what the outcomes are for the student that goes to this institution. Did they get what they needed from us, and has it improved their lives? And the vast majority, the answer is "yes." (Dean of Instruction, Pima College)

Outcomes, ultimately, are employment for the student or upgraded skills within an occupational or vocational field. What students need, according to this administrator, are economic improvements.

In the second example, I ask the vice president of Wake Technical Community College if there has been change to the goals and operations of continuing education programs over the past two decades. His response indicates that the change has been fundamental; he notes, too, that this alteration to training and retraining was his personal goal as well. ("Q" indicates the question asked and "A" the response.)

Q: Do you see a shift in continuing education programming over the past twenty years from more self-development, personal development, and basic development to more occupational?

A: Absolutely, no doubt about it, tremendous shift. It should have been that way all along. . . . The community service type[s] of things . . . were so prominent in the seventies. There has been a shift, a complete shift, not complete but a vast shift, to the more training, retraining, employment types of things. (Vice President, Continuing Education, Wake Technical Community College)

These two views—one from an administrator in Arizona and one from an administrator in North Carolina—are unequivocal about the orientation of continuing education and the biases and assumptions of those who manage these program areas.

In the third example, the president of the North Carolina State Community College system amplifies the workforce and economic role of colleges. His view indicates that continuing education is not only vocational and occupational training, but also specific to a state's economy. His blending of opportunities for students with market forces and responsiveness to business is also a connection of the traditional orientation of community colleges to social mobility and equity for students.

That [social mobility and equity] is a major mission. This system was created in a different format in 1958, to do just customized training for industry . . . and then five years later it became a comprehensive system and it included other programs. But thirty-five years ago or forty years ago, we were created to help North Carolina through a major economic transition: from an agrarian economy to a manufacturing economy. We are now going through an equally wrenching transition from traditional manufacturing to knowledge-based manufacturing. Tobacco, textiles, and furniture have gone or are going. Textiles are gone; tobacco will gradually wind down; and furniture is going; and [it is] every bit as traumatic for those workers as it was for farm workers in the fifties and early sixties, when economization and other factors made them lose their jobs. So, I think we are the key to this economic transition and this social change in North Carolina. Nobody can do it but us, in my opinion: too massive for the universities to undertake. For an overwhelming majority of these displaced workers they simply do not have the academic background to benefit from a four-year program. But because of our excellence in remediation and assessment, building on success, we can take these people who have not been in a classroom in twenty-seven years and give them a new skill and a new hope in a way we believe nobody else can. So, that is our mission and has been from day one.

Almost all students, this executive leader notes, are workers in a new labor force for the state.

The fourth example is from the state's chief executive officer for community colleges in California.

I want to see the colleges much more aligned with business leadership in the state . . . [with] all sorts of cross-fertilization between the business community helping the colleges with both politics and fundraising, but also then the colleges helping the businesses, the economic family . . . helping recruit and retain business, helping better trained workers.

The goal is to improve the social and financial stature of the community college and ultimately to secure jobs for students. This college-business nexus, while of benefit to students economically, marries the community college to business and industry. Community colleges serve as proxies for corporations and businesses in their preparation of employees.

A second set of issues is evident in the responses of administrators on institutional practices at two community colleges in two states—Texas and Washington. These issues concern the educational function of community colleges, specifically the unit of continuing education's role in student development. The first part pertains to the blurring of boundaries between traditional technical education, which serves students in their educational development, and workforce training through continuing education, which serves business and industry.

I think another thing we've tried to do, and it's interesting you said continuing ed, because and, you know this, that blurring of the line between

continuing ed and particularly technical education, where in many cases—
and one of our colleges uses as the model—almost every new technical program they put in they start as a continuing ed program because there are
some faculty issues you don't have to deal with. You don't have to have a full-
time faculty member. And it really is a way of putting your foot in the water
and testing it out and seeing if it's going to be successful. (Dallas
Community College District, District Associate Vice Chancellor for
Educational Affairs)

The second part reflects an administrator's critique of her own institution's practices in educational advancement for lower achieving students. That is, the students meet course or program expectations, but their educational development
may be arrested.

I think what we probably don't do, and what we talk a lot about is, we don't
create opportunities and really facilitate the movement of those students
from one level to another very well. So . . . I mean we're starting that, but
a student could come in at ABE [Adult Basic Education], GED [General
Educational Development] and it could be that they come in and they do
that, but we have not really done a very good job of helping move them
into a particular career program or even into academic transfer. We've not
facilitated that movement. We've enabled them, we've prepared them
through ABE, GED, but we've not really, we don't, we do it with a small
group, but I would say overall we don't do a very good job of that. (Vice
President Workforce Development, Edmonds Community College)

The administrators noted above, with the exception of the vice president at
Edmonds Community College, frame the mission and function of the community college as largely economic, with student preparation for employment
through skills development as outcomes, whether as goals or as achievements.
For them, continuing education is clearly workforce training or retraining. For
the vice president at Edmonds, there is acknowledgment that the outcomes are
training or retraining, but some hesitancy to laud these efforts as educational
achievements.

There may be conflict between community college goals for student academic development and the state's need for a trained workforce. On the one
hand, practitioners address student needs and education and training requirements. On the other hand, practitioners frame lifelong learning as an economic
imperative, with a focus upon workforce development, largely for private business and industry. Where lifelong learning is associated with workforce development for the private sector, the ascendancy of lifelong learning in community
colleges arises from and reinforces a "consumer-centric model" that places the
institution in an economically competitive environment. What this means is
that community colleges function in line with the priorities of the state,[41] which
are increasingly patterned after a neoliberal ideology. Because the state is both a

critical fiscal resource and a legal authority for public community colleges, the state's influence in considerable.

College administrators are cognizant of their business and industry and workforce preparation orientation of programs and instruction. However, they do indicate that student outcomes—primarily in the form of jobs or skill acquisition for jobs—are of particular concern. Yet students are viewed as economic entities, with primarily or solely economic needs. Thus conflict is avoided because the prevailing view is that what is good for business and industry is good for students. Lost or missing in this discussion, however, is that justification for a community college education that is compatible with earlier virtues of lifelong learning—the acquisition of new knowledge that pertains to all facets of life in a complex postindustrial society.[42] Arguably, students at the lower levels of academic achievement and attainment—such as those in Basic Skills programs—are less likely to flourish and more vulnerable in a postindustrial society than those at the higher levels of attainment—such as those in academic transfer and high technology programs.

In large part, then, well-defined continuing education units at community colleges have economic functions—they generate revenues for the institution, meet the workforce needs of employers, and prepare, retrain, or upgrade skills of workers. Profit for the institution is a central function.

> [We] have only probably contributed $100,000 each year to the general fund. . . . My hope is that by this coming summer we are able to increase and contribute $150,000, and then the following year contribute $200,000. We are just starting a very large marketing program. Again [this] is not typically done in the community college system. We spent the last couple months developing it, part of it a business plan, marketing plan [to] identify 3,000 companies that are not in our district. We will send them different postcards every two weeks—thirteen mailers. [We] hired a telemarketing firm that will call 500 companies every two weeks for us to go out to do that contract. (Dean of Community Advancement, El Camino College)

There is considerable emphasis upon revenue generation for community colleges, especially those in urban and suburban areas where student populations and the local population are growing. Contract training is among the major revenue-generating activities, yet colleges' "profits" are not impressive. Take, for example, the entire Dallas Community College District, with a reported 62,706 credit students and 19, 935 noncredit students in the spring 2005 period, with seven separately accredited colleges. Recognizing that contract training is not the "cash cow" once expected, the colleges continue to find other ways to generate revenues.

> Overall in our district, especially the last couple of years because of the tight economic situation, we have certainly not generated the revenue that

we did, say, three or four years ago. . . . I'm not talking profit; I'm just talking about gross, gross income. About four years ago we were grossing
about seven million [dollars] for the entire district in contract training
alone. This last year we grossed just a little over five million [dollars]. So
you can see that, you know, [in] tight economic times there was a drop of
about $2 million gross. But then, of course, you have to subtract from that
gross what your direct costs are and indirect costs, and so on. You know,
quite honestly in some ways, in some ways I think our system is beginning
to recognize yes, contract training can be used as an income generator, but
it may not be as big of an income generator as [we] once thought it could
be. . . . [T]here's even more effort being looked at for other alternate
sources for income: some of that being through grants and resource development kinds of things. (Director of Workforce Education, Dallas County
Community College District)

According to the 2004 audit, total district operating revenues of Dallas
Community College District equaled approximately $293,600,000, and of these
revenues $5,168,000 were gained from nongovernmental contracts—less than 2
percent—whereas tuition and fees generated approximately 15 percent. The
overwhelming revenue for the colleges came from government—federal government and state government grants and state government appropriations.
Thus, the focus and attention to revenue generation seems misguided even on
economic counts. Much of the focus upon contract training and revenue generation is for the improvement of private business, not necessarily for the public
good, unless we think of the training of city workers as the public good. At
Piedmont Virginia Community College in Charlottesville, Virginia, the director
of workforce services expresses a view suggesting that the private sector's priorities are those of the college's.

[Contract training] seems to be growing. It's . . . favorable to them, the
companies, because it's geared toward their specific company, and we can
customize it to some of their specific documents or their scenarios. . . .
[W]hen they role-play they can use their specific company. So it's worked
fairly well. We work . . . with Clockner. We've done leadership programs
with their managers. [The] city of Charlottesville, we're doing . . . their
computer training, and doing leadership with them also, with their managers. . . . Tiger Fuel, we're doing all their orientation and customer service training. Guarantee Bank, we're doing their customer service and
doing some coaching with their managers. G. Fanneck—we're just doing
PC repair courses for them. Technicolor, we did some electrical safety
courses for them before they left town. . . . [W]e're getting ready to work
with Wal-Mart on some Spanish for their new hire managers. The
biggest . . . thing we did last year, last spring, we met with what we call
business education roundtables or BERTS. We had nine business clusters.
We brought in business folks from each of those clusters, asked them what

their needs were, and we are now in the response mode of telling them what we've done to meet those needs. . . . From that initiative, we started truck driving school, commercial driving school. We also started . . . workplace essential skills. That one includes basic math, writing skills, along with employability skills, just attitudes and things of that nature. That's one that we heard commonly among all the groups. They were saying that we need people that will show up and have a good work ethic, will do the job they're asked to do and continue to do it. . . . [A]ll of them were saying that was an issue with folks that they're getting. . . . Spanish for the workplace for the adult managers . . . we created [that]. So we've tried to react to what they said, and we're developing programs, in general, that will meet that, and then we'll also, again, customize if we're asked to or go outside the needs of that.

All of this energy leads to a profit of $50,000 per year for the college to use as it sees fit. This is a public college supported largely for its operations with funds from taxpayers, and its continuing education unit is training employees at businesses and industries—including corporations such as Wal-Mart—to help increase their profits. The published mission statement of the college, while emphasizing student educational development, also notes that workforce development is central, and that includes "meeting the . . . needs of employers." In this context, then, Piedmont Virginia Community College serves as an instrument not only of the state but also of business and industry.

> Piedmont Virginia Community College promotes student success through excellent educational programs and services that are accessible and affordable.
>
> The college is a comprehensive, public, associate degree-granting institution. As part of the Virginia Community College System, Piedmont Virginia Community College serves the City of Charlottesville and the counties of Albemarle, Buckingham, Fluvanna, Greene, Louisa, and Nelson.
>
> College transfer and workforce development are the core of the college's mission. Challenging coursework and a full range of support services are provided for students in both college transfer and workforce development programs. The first two years of baccalaureate study prepare students for success at four-year colleges and universities. Workforce development programs prepare students for successful careers and promote a skilled regional workforce by meeting the training and educational needs of employers. Programs and services in developmental education, general education, community service, and lifelong learning support and enhance the mission core and prepare students for success in life. (Mission statement, Piedmont Virginia Community College, adopted 2001)

As society reorganizes and restructures, educational institutions take on both reflective and facilitating roles. As social attitudes and values shift, educational institutions both reflect and respond to these shifts. One the one hand, we have a progressive "modern" society moving toward an advanced and enlightened system; on the other hand, we have a postmodern society grasping for both solidity and the ability to accept and cope with flux.

College students continue to behave in traditional, modern ways, as do institutions of higher education: Take courses, fulfill the requirements, earn credits, move on to the next level, or follow the sequence. But, the lives of these college students are not traditional or modern, even though there is a façade of continuity. At El Camino Community College in California, students are multiethnic, multiracial; they have complex lives, working lives, family or parental lives; they are immigrants or children of immigrants. They are not from highly educated families, or from families with wealth, and they have no wealth themselves. In addition, there is a second grouping of students who are not students as their primary role. They do not think of themselves as students, and several who are owners of small businesses use the college to secure a loan or develop a business plan. They are neither students nor customers: They are shopkeepers, not students; employers, not students. Others within this grouping take medical terminology courses to upgrade their work as receptionists at a medical office; and others take a medical billing program that carries no college credit but gives them a national certificate so that they can access jobs that pay them $12 per hour. Some have few skills upon entry to college; others have graduate degrees. Some want to start their own company; others just want company work.

We have arrived at a society of lifelong learning, not because we have evolved to accept that learning *per se* is a defining or developmental characteristic of being human or because we have much to learn to be human and realize our capacities, but because the new economy, the knowledge economy, the globalized competitive economy requires us to serve its masters and to adjust to changing technical requirements. Lifelong learning is an adaptation to the corporation, to the company, to the institution, to the state. Lifelong learning is what the corporation must have for its workers or it will not be competitive. Some learning is about skills—computing, mechanical communication—and some is about merging oneself with the organization: accepting the goals and values; adopting organizational language and behaviors; and selling the organization and its products or services to others.[43] The question for community college educators, from a Rawlsian perspective of justice, is: Are the disadvantaged—those who have lower social and economic standing than the privileged in society—advantaged?

Perhaps community college administrators have little alternative but to articulate and promote a neoliberal ideology for community colleges, given their institutions' subordinate role and dependent position in relation to the state and to business and industry. Corporate values and behaviors, such as accountability, with a decidedly economic orientation, are on the ascendant in higher and adult education,[44] and these values can be seen in organizational orientations to the economic marketplace and in curricula that places high priority on employment skills. These skills—communication, teamwork, and critical thinking[45]—are reified and then identified with what business wants: a competitive workforce. This

workforce, which is a product of lifelong learning, is designed to fit the new corporate culture of high productivity, global competitiveness, and continuous
workplace change that has become the norm for advanced industrial states.[46]

The Exclusionary Effects of Lifelong Learning

Community colleges as institutions have for several decades prided themselves
on offering an "open door" or "open access,"[47] but organizational behaviors of
community colleges that are affiliated with neoliberal concepts of lifelong learning are inadvertently excluding populations from education and training. Skills
required by corporations and business to compete globally are not so much the
old vocational ones of the craftsperson or journeyman or blue-collar worker, but
rather employability skills that include attitudes favorable to the employer and
high-level technical skills, often technologically advanced.[48] Lifelong learning as
an adaptive strategy of organizations privileges core workers and ignores
peripheral workers, such as service workers.[49] Learning for global competitiveness is viewed as an investment; developmental or remedial learning, as well as
nonvocational education, is viewed as consumption and not supportable by the
state or the public within a neoliberal philosophy. Thus, low socioeconomic
populations are twice excluded: They are excluded from avenues to vocations
and from public support for their educational upgrading. When they do participate in college education in academic development courses, they may experience a level of education that is not comparable to that provided in other sectors
of the institution. In North Carolina, basic education students are funded by the
state at a lesser percentage than students in degree programs, which often means
the resources that institutions provide them are not compatible with economic
or social mobility in a postindustrial economy. Evelyn teaches basic education
at Johnston Community College in North Carolina. While meeting the developmental and emotional needs of her students, she is unlikely to provide them
with the technological and employment skills that corporate employers
demand. Evelyn is paid under $400/week for her efforts; she is contingent labor
and detached from the mainstream faculty of the institution.

> I teach each day from eight o'clock in the morning until . . . noon. And I
> teach five subjects: language arts, and reading, language arts, writing,
> parts 1 and 2. I teach science, social studies, and math, parts 1 and 2. And,
> of course, I have students that come from all around and we start the day
> off with our objectives and they work very hard and I challenge them to
> do that. I challenge them to set goals for themselves, and of course [to] be
> able to obtain those goals that they would be obtainable. . . . They've been
> out of school; it is a challenge for them, but it's amazing how they, as I said,
> set their goals. This is something they've wanted to do for a long time, and
> they're there to do it. . . . I ask many things of them, but mainly to come
> to class; also to set academic goals for themselves and personal goals, and
> hopefully work goals because some of them do work and go to school,

too. . . . I am a person that believes in setting goals, individual goals. And these goals, there is a link between goals and values, attitudes, and steps. So if you design a plan, you have a series of steps that you can use. One step is a small step, or an action, and it links with the next step. . . . [T]his is one way they can participate in their own motivation to succeed if they do something, and they succeed in doing it, and they accomplish certain skills and concepts. Then they feel like that they are progressing, because they have to have that feeling to stay with you. Or else, if they don't have that feeling . . . you don't see them in about a month or so, and then they have to be dropped. They just don't produce, so they have to have this feeling of achievement or accomplishment, however little or large, or big or small. They have to feel like they are succeeding in their goals. (Evelyn, Faculty, Johnston Community College)

There is no specific skills development and no use of technology; there is little evidence that developmental students are going to be participants in the new economy unless as service workers. For the federally supported students in welfare-to-work programs, the goal is not employability in the knowledge economy but service sector work as quickly as possible. The dean at Wake Technical Community College in North Carolina laments this condition. She notes as well that work opportunities are in the long run illusory.

It's get to work quick, not, "Let's train you in a trade so that you are going to be able to support your family. But we want to get you off the road and we want to get them out as quickly as possible." They [policymakers and legislators] want a quick fix, but they [students] find out they still cannot support their family, then what are they working for. There is a group [with] limited social services benefits, like, after they get a job I think they can stay there for a year; after that year is up, they are going to have to quit work, because they can't afford to work and pay for someone to take care of their kids. They don't make enough to work and keep their kids in day care. So, you know these statistics that you hear, how many are off the welfare road, but what is going to happen to them two years down the road, when they run out of these supplemental benefits. (Dean, Adult Basic Education, Wake Technical Community College)

What happens to disadvantaged students under this imperative is a matter often ignored or rationalized within an economic human capital context. Globalization theory applied to higher education suggests that there are clear "winner" and "loser" programs and units.[50] There is evidence as well that student populations that are disadvantaged in higher education are those associated with the programs and units that fall under the "loser" categories.[51] These programs are affiliated with or comprise continuing education: Special Education, Basic Education, English as a second language, short-term certificate programs, GED, Adult High School, and a host of retraining programs.

Students within these programs who are disadvantaged students because of economic status, educational background, disability, and the like are excluded from the traditional social and economic benefits of higher education, excluded from participating in the knowledge economy except as ancillary workers and often contingent workers. Against a backdrop of government and institutional claims about access to education and participation in society there is inequitable treatment of these classes of disadvantaged populations.[52]

Notes

1. Richard Bagnall, "Lifelong Learning and the Limitations of Economic Determinism," *International Journal of Lifelong Education* 19, no. 1 (2000): 20–35; Richard Edwards, "Lifelong Learning and a 'New Age' at Work," in *Lifelong Learning and Continuing Education: What Is a Learning Society?* ed. Paul Oliver (Brookfield, VT: Ashgate, 1999), 31–45; Jack Fuller, *Continuing Education and the Community College* (Chicago: Nelson-Hall, 1979); Ken Meier, *The Community College Mission: History and Theory* (unpublished manuscript, Bakersfield, CA: 2004); Robin Usher, "Identity, Risk, and Lifelong Learning," in *Lifelong Learning and Continuing Education: What Is a Learning Society?* ed. Paul Oliver, 65–82 (Brookfield, VT: Ashgate, 1999); Shirley Walters and Kathy Watters, "Lifelong Learning, Higher Education, and Active Citizenship: From Rhetoric to Action," *International Journal of Lifelong Education* 20, no. 6 (2001): 471–78.
2. Arthur Cohen and Florence Brawer, *The American Community College* (San Francisco: Jossey-Bass, 2003).
3. John S. Levin, *Globalizing the Community College: Strategies for Change in the Twenty-First Century* (New York: Palgrave, 2001); Levin, "Student Markets: The Business Culture of the Community College" (paper presented at the annual meeting of the Association for the Study of Higher Education, Sacramento, CA, 2002); Levin, "The Business Culture of the Community College: Students as Consumers; Students as Commodities," in "Arenas of Entrepreneurship: Where Nonprofit and For-profit Institutions Compete," ed. Bruce Pusser, special issue, *New Directions for Higher Education* 129 (2005): 11–26.
4. Thomas R. Bailey and Irina E. Averianova, "Multiple Missions of Community Colleges: Conflicting or Complementary" (occasional paper, Community College Research Center, Teachers College, New York: 1998); Quentin Bogart, "The Community College Mission," in *A Handbook on the Community College in America*, ed. George Baker, 60–73 (Westport, CT: Greenwood, 1994); Cohen and Brawer, *The American Community College*; Kevin Dougherty, *The Contradictory College* (Albany: State University of New York Press, 1994).
5. Levin, *Globalizing the Community College*; Levin, "Business Culture."
6. Bogart, "The Community College Mission"; Steven Brint, "Few Remaining Dreams: Community Colleges since 1985," *The Annals of the American Academy of Political and Social Sciences* (March 2003): 16–37; Brint and Jerome Karabel, *The Diverted Dream: Community Colleges and the Promise of Educational Opportunity in America, 1900–1985* (New York: Oxford University Press, 1989); John Dennison and Paul Gallagher, *Canada's Community Colleges* (Vancouver: University of British Columbia Press, 1986); W. Norton Grubb, *Honored but Invisible: An Inside Look at Teaching in Community Colleges* (New York: Routledge, 1999).

7. Lisa Hudson, "Demographic Attainment Trends in Postsecondary Education," in *The Knowledge Economy and Postsecondary Education*, ed. P. A. Graham and N. Stacey, 13–54 (Washington, DC: National Academy Press, 2002).

8. Bagnall, "Lifelong Learning," 25–27.

9. Patricia Gouthro, "Education for Sale: At What Cost? Lifelong Learning and the Marketplace," *International Journal of Lifelong Education* 21, no. 4 (2002): 334–46; David Shupe, "Productivity, Quality, and Accountability in Higher Education," *Journal of Continuing Higher Education Winter* (1999): 2–13; Kenneth Wain, "The Learning Society: Postmodern Politics," *International Journal of Lifelong Education* 19, no. 1 (2000): 36–53.

10. Usher, "Identity, Risk, and Lifelong Learning."

11. Edwards, "Lifelong Learning."

12. Grubb, Noreen Badway, and Denise Bell, "Community Colleges and the Equity Agenda: The Potential of Noncredit Education," *The Annals of the American Academy of Political and Social Science* (March 2003): 218–40.

13. Kent A. Phillippe and Leila Sullivan Gonzalez, *National Profile of Community Colleges: Trends and Statistics*, 4th ed. (Washington, DC: American Association of Community Colleges, 2005).

14. Michael Apple, "Comparing Neoliberal Projects and Inequality in Education," *Comparative Education* 37, no. 4 (2001): 409–23.

15. Dougherty and Marianne Bakia, "Community Colleges and Contract Training: Content, Origins, and Impact," *Teachers College Record* 102, no. 1 (2000): 197–243; Steven Lee Johnson, "Organizational Structures and the Performance of Contract Training Operations in American Community Colleges" (unpublished doctoral dissertation, University of Texas at Austin, 1995).

16. Cohen and Brawer, *The American Community College*; Dougherty and Bakia, "Community Colleges"; Government Accountability Office, "Public Community Colleges and Technical Schools (No. GAO-05-04)" (Washington, DC: United States Government Accountability Office, 2004).

17. GAO, "Public Community Colleges," 4.

18. Davis Jenkins and Katherine Boswell, "State Policies on Community College Workforce Development" (Denver, CO: Education Commission of the States, Center for Community College Policy, 2002).

19. Dougherty and Bakia, "Community Colleges."

20. I am indebted to Tom Collins, "*Contract Training as a Revenue Source for Community Colleges,*" (unpublished manuscript, Raleigh, NC: 2005).

21. Bagnall, "Lifelong Learning"; Ken Meier, "Social and Educational Origins of the Community College Movement: 1930–1945" (unpublished manuscript, Bakersfield, CA: 2004).

22. Anthony P. Carnavele and Donna M. Desrochers, "Community Colleges in the New Economy," *Community College Journal* 67, no. 5 (April/May 1997): 26–33; Levin, *Globalizing the Community College.*

23. Catherine Casey, *Work, Society, and Self: After Industrialism* (New York: Routledge, 1995).

24. D. Franklin Ayers, "Discursive Manifestations of Neoliberal Ideology in Community College Mission Statements: A Critical Discourse Analysis" (unpublished manuscript, University of North Carolina at Greensboro, 2004).

25. Jim Gallacher et al., "Learning Careers and the Social Space: Exploring the Fragile Identities of Adult Returners in the New Further Education," *International Journal of Lifelong Education* 21, no. 6 (2002): 493–509.

26. Gouthro, "Education for Sale."
27. Bagnall, "Lifelong learning."
28. Michael Skolnik, "The Virtual University and the Professoriate," in *The University in Transformation: Global Perspective on the Futures of the University*, ed. Sohail Inayatullah and Jennifer Gidley, 55–67 (Westport, CT: Bergin and Garvey, 2000).
29. Brian Pusser, *Burning Down the House: Politics, Governance, and Affirmative Action at the University of California* (Albany: State University of New York Press, 2004).
30. Simon Marginson and Mark Considine, *The Enterprise University: Power, Governance, and Reinvention in Australia* (New York: Cambridge University Press, 2000); Sheila Slaughter and Gary Rhoades, *Academic Capitalism and the New Economy: Markets, State, and Higher Education* (Baltimore: Johns Hopkins University Press, 2004).
31. Arthur Cohen and Florence Brawer, *The American Community College*, 3rd ed. (San Francisco: Jossey-Bass, 1996); Levin, Susan Kater, and Richard Wagoner, *Community College Faculty: At Work in the New Economy* (New York: Palgrave Macmillan, 2006); Dennis McGrath and Martin Spear, *The Academic Crisis of the Community College* (Albany: State University of New York Press, 1991).
32. Bailey and Vanessa Smith Morest, *The Organizational Efficiency of Multiple Missions for Community Colleges* (New York: Teachers College, Columbia University, 2004).
33. *The Chronicle of Higher Education, Almanac Issue 2004–5* (Washington, DC: 2004): 16.
34. Phillippe and Gonzalez, *National Profile of Community Colleges*.
35. Dougherty and Bakia, "Community Colleges"; Dougherty and Bakia, "The New Economic Role of the Community College: Origins and Prospects" (occasional paper, Community College Research Center, Teachers College, New York, June 1998); Levin, "The Business Culture of the Community College."
36. Bailey and Morest, "Organizational Efficiency"; Levin, "The Revised Institution: The Community College Mission at the End of the Twentieth Century," *Community College Review* 28, no. 2 (2000): 1–25; Levin, "The Business Culture of the Community College."
37. Levin, Kater, and Wagoner, *Community College Faculty*.
38. D. Franklin Ayers, "Neoliberal Ideology in Community College Mission Statements: A Critical Discourse Analysis," *The Review of Higher Education* 28, no. 4 (2005): 527–49.
39. Casey, *Work, Society, and Self*.
40. Ibid., 191.
41. Arthur Cohen, "Governmental Policies Affecting Community Colleges: A Historical Perspective," in *Community Colleges: Policy in the Future Context*, ed. Susan Twombly, 3–22 (Westport, CT: Ablex, 2001); Dougherty, *The Contradictory College*; Levin, "Public Policy, Community Colleges, and the Path to Globalization," *Higher Education* 42, no. 2 (2001): 237–62.
42. Bagnall, "Lifelong Learning"; Gouthro, "Education for Sale."
43. Casey, *Work, Society, and Self*.
44. Patrick Fitzsimons, "Changing Conceptions of Globalization: Changing Concepts of Education," *Educational Theory* 50, no. 4 (2000): 505–21; Bill Readings, *The University in Ruins* (Cambridge, MA: Harvard University Press, 1997); Slaughter and Rhoades, "The Neoliberal University," *New Labor Forum* (Spring/Summer 2000): 73–79; Carlos A. Torres and Daniel Schugurensky, "The Political Economy of Higher Education in the Era of Neoliberal Globalization: Latin America in Comparative Perspective," *Higher Education* 43 (2002): 429–55; Anthony P. Welch,

"Globalisation, Postmodernity, and the State: Comparative Education Facing the Third Millennium," *Comparative Education* 37, no. 4 (2000): 475–92.

45. Carnavele and Desrochers, "Community Colleges"; Terry O'Banion, *The Learning College for the Twenty-First Century* (Phoenix, AZ: American Council on Education and the Oryx Press, 1997).

46. Casey, *Work, Society, and Self.*

47. Brint, "Few Remaining Dreams"; Burton Clark, *The Open Door College: A Case Study* (New York: McGraw-Hill, 1960); Cohen and Brawer, *The American Community College*; Alicia C. Dowd, "From Access to Outcome Equity: Revitalizing the Democratic Mission of the Community College," in "Community Colleges: New Environments, New Directions," ed. Kathleen Shaw and Jerry Jacobs, special issue, *The Annals of the American Academy of Political and Social Science* 586, no. 1 (2003): 92–119; Roueche, Eileen E. Ely, and Roueche, *In Pursuit of Excellence: The Community College of Denver* (Washington, DC: Community College Press, 2001).

48. Edwards, "Lifelong Learning"; Wain, "The Learning Society"; Roger Waldinger and Michael I. Lichter, *How the Other Half Works: Immigration and the Social Organization of Labor* (Berkeley: University of California Press, 2003).

49. Bagnall, "Lifelong Learning"; Wain, "The Learning Society."

50. Levin, *Globalizing the Community College*; Slaughter and Larry Leslie, *Academic Capitalism, Politics, Policies, and the Entrepreneurial University* (Baltimore: Johns Hopkins University Press, 1997).

51. Marion Bowl, *Nontraditional Entrants to Higher Education* (Stoke on Trent, UK: Trentham Books, 2003); Penelope E. Herideen, *Policy, Pedagogy, and Social Inequality: Community College Student Realities in Postindustrial America* (Westport, CT: Bergin and Garvey, 1998); Shaw et al., "Putting Poor People to Work: How the Work-First Ideology Eroded College Access for the Poor" (unpublished manuscript, 2005).

52. Bowl, *Nontraditional Entrants to Higher Education.*

Chapter 8

Justice and New Nontraditional Community College Students

In this chapter, I reiterate the problem that propelled this book and I suggest solutions that institutions and policymakers can consider in addressing the problem and in granting justice to nontraditional students. As an institution that historically has claimed the moral high ground for educating underserved populations—including those populations who are on the economic periphery of society—the community college bears a particular responsibility for remedying unjust conditions for disadvantaged populations. For the large part of the twentieth century, community colleges endeavored to find a legitimate and respectable role in the educational landscape, including as a paraprofessional workforce trainer, a junior college, and even an applied baccalaureate site, as well as a community service organization.[1] The end of the twentieth century and the beginning of the twenty-first saw the emergence of an institution closely identified with a global economy and directed by "new managerialism,"[2] consistent with other higher educational institutions. Yet, there are differences: The public community college is clearly an extension of the state, legally under the authority of state legislatures; and because of its local orientation and training function, the institution is dependent upon local needs and demographics. Both its students and its faculty have characteristics that are not typical of four-year colleges and universities, and its curriculum, with a comprehensive array of programs from developmental education to high-tech training and university transfer courses, separate the institution from others in the postsecondary field.[3]

Its institutional characteristics, history, and particularly its student population suggest that the community college has a social responsibility for advantaging the disadvantaged. While economic practices such as generating revenue and pursuing greater levels of efficiency are difficult for community college leaders to disregard, given the level of funding provided by the states for their increasing enrollments and services, these practices have become more central to the behaviors of the institution, suggesting they are ends in themselves.[4] Legislative interventions and oversight, while justified as serving the public interest and conserving fiscal resources, tend to emphasize the economic behaviors of institutions and ignore both educational outcomes and vulnerable populations. For example, the reactionary stance of legislators to undocumented immigrants not only penalizes a vulnerable group but also restricts the long-term knowledge capacity and economic vitality of the states. A precious resource—the human potential for highly productive work—is risked as large numbers of undocumented immigrants, faced with exorbitant costs for postsecondary education, do not attend these learning institutions, although they could benefit from attendance. For all vulnerable populations and those who are disadvantaged, community college leaders and policymakers at the state and national levels have an obligation to remedy injustices.

The Condition of New Nontraditional Students

Community college students are tracked by their status as credit students; if they are in programs categorized within their state as noncredit, they are not reported in data sets and are ignored in scholarly examinations of students. Within the credit framework, students in community colleges are a heterogeneous group: 33 percent are classified as minority; 45 percent are first-generation students in college; 35 percent consider themselves to be workers who also attend college (40.8 percent work full-time while attending college); over 50 percent do not enter college directly from high school; 17.2 percent are single parents; and 66.1 percent attend college part-time.[5] They are considerably different from their four-year college counterparts: They have less time and less money to devote to college. Large numbers of students in noncredit offerings—such as basic skills, adult high school, and English as a second language (ESL)—have fewer cultural, educational, and social assets than their credit counterparts. Students who are not traditional college students have goals and preoccupations that are different from those who we find in accounts of college students, such as Marcia Baxter Magolda's *Making Their Own Way: Narrative for Transforming Higher Education to Promote Self-Development*, Dorothy C. Holland and Margaret A. Eisenhart's *Educated in Romance: Women, Achievement, and College Culture*, Michael Moffat's *Coming of Age in New Jersey: College and American Culture*, and Rebekah Nathan's *My Freshman Year: What a Professor Learned by Becoming a Student*.[6] Students reported in these works reflect problems and concerns such as finding a sexual partner while in college, getting by (i.e., attaining a passing grade with the least amount of work necessary) in college courses, and surviving the passage from adolescence to adulthood. Tom Wolfe's *I Am Charlotte Simmons*, although a work of fiction, satirizes the behaviors of traditional colleges students in research universities and reinforces the observations of social scientists who have done ethnography to capture college students' experiences.[7]

Nontraditional students, and particularly disadvantaged students, draw upon different experiences—such as immigrant backgrounds—and base their decisions and choice making on their social and domestic conditions. O., a female student from Nigeria, has lived in the Chicago area for eight years, attends Truman College to gain English language skills, and has a goal of beginning a business. Her family is fragmented, with three of her children living in Nigeria;

her educational background is limited: She did not complete secondary school, and English is not her primary language.

> I am just trying to learn English. It is hard for me, but I am doing it . . . English and math. . . . I am trying to make it so I can be different than before. That is my goal. . . . I would like to do business. . . . I didn't finish my high school because of my children. Me and my husband, we tried to pick up children [from school]. So, I dropped my school . . . [while] taking care of children. . . . I have five [children]. . . . Two of them [are] here, they came to America. Two years now. I have two here in America now. The rest are in Nigeria. . . . I want someone to push [me], someone to help me read, read, and read. Like, if I don't know, "Can you please help me?" "This spelling, can you please help me?" Just someone who can help push me through everything and help me finish quickly. It could even be somebody that asks if you have been going to class; helps you if you don't know spelling. That helps you know what to look for, so you don't look like a fool.

O. is not a student attempting to fit in to dormitory life, a sorority, or even the freshman-year experience. Her "life world" and her social world are not college. Even for those nontraditional students who are highly dependent upon college for their present sense of place and personal identity—such as the special education students at Johnston Community College in North Carolina or the disabled students at Bakersfield College who participate in a college course called Tools for College Survival, or Mona at Community College of Denver, who is an epileptic and for whom the world in general is a fearful place—college is only one factor in their "life world" and in their choices about their future circumstances.

> I'm disabled; I have epilepsy. I was scared to go to college because of my sickness, my health. And that is why I don't have a job, because of my seizures, and I say to myself, "How can I live?" I have two girls. I'm not married. I say to myself, "How can I live?" Yesterday I just made [a necessary purchase and] there went all my SSI checks. I got thirty-two cents back. I said to myself, "How can a girl live on thirty-two cents a month?" (Mona, student, Community College of Denver)

This is not to say that Mona is without skill or potential; her experience as an interpreter will serve her in her new aspiration as a medical assistant, where her Spanish language skills will aid patients.

> I worked at the elementary schools for five years [as an interpreter]. One year they would send me to one school and the next year to another school, and most of the students used to come to my desk and ask me in Spanish, "What is the teacher saying?" Because she was speaking in English, they did not understand a word. And it did not confuse me because I've known Spanish all my life. So I would tell the child, the student, to sit down, and

I will, let me listen for a second and I will interpret it back to you. And the child was very pleased with it every day, and when I was going to leave, the children, the students who needed my help, came up to me and didn't want me to leave. And that pleased me quite a bit. The reason why I did leave was [because] my father had his fourth stroke and you just don't want to take it out on the children, get on their nerves for it. So when my father was controlled, OK I tried to go back to the schools. Because before I started working for them I even volunteered. I live in Englewood, Colorado, and my daughters went to the school in Englewood and they needed interpretation there, too, [and I] volunteered. At [that time] I wasn't thinking about working. So I would go and interpret for them. (Mona, student, Community College of Denver)

The reasons for students participating in postsecondary education are multi-dimensional: Decisions for nontraditional students are complex, ranging from costs and time availability to self-perceptions and personal histories, as well as their physical, psychological, and mental states.[8]

In this vein, then, philosophies and ideologies that emphasize self-help or individual responsibility ring hollow: Not only do they lack compassion, but they also make false assumptions about equal opportunity. Community colleges and their students are at the educational vortex of neoliberal policies that favor individual responsibility, the diminution of the welfare state, and the privileging of elites. These policies work against disadvantaged students.

The Politics of College Education

If education and training are tied to fiscal resources, then these are political expressions first within the institution, second at the state level, and third at the level of power and influence beyond the state. Access for students is a political matter, whether access is about the provision of specific programming or about costs to students. And the political must come down to values and ideology. For example, state policy in California, as expressed through legislation on the mission and goals of community colleges, targets economic development as one of the principal purposes of community colleges, yet funding from the state for job preparation and vocational education does not conform to either the intent or even to the rhetoric of workforce development. As there is essentially no differential in allocations to community colleges based upon programming, the incentive is for colleges to offer what is most cost-effective. This behavior leads to an abundance of general education and transfer courses that cost the least to provide within the college curriculum. As an expression of values, government funding policy serves the demand of the population—academic courses—even though university transfer itself is a low result outcome. For those in the transfer stream, 20 percent actually transfer.[9] Out of 15,000 credit students at Bakersfield College in California, 1,000 transfer annually to a university. This behavior of government funding suggests that the community college is not valued for its workforce or economic development but is instead valued for its ability to house students and delay their entry to either the labor market or the employment lines, or else ensure their underemployment as part-time marginal workers.

College administrators and faculty work toward resisting this condition and endeavor to make significant change to the lives of students. At Bakersfield College, college officials target underrepresented populations and at-risk students and provide service and programs such as cooperative education and distance education to carry students to employment as early childhood educators, registered nurses, and teachers. For the majority of students, these services and programs are their only opportunity to gain an education that will lead to both personally well-regarded and economically sustainable employment. For administrators and faculty who support and guide these students, the actions are part of a personal value system and agenda for the betterment of these students. At the same time, administrators must make less palatable decisions and act to reduce the college's labor force[10] by approximately 20 percent to cope with expected shortfalls from state government and serve layoff notices, a situation that was evident in 2003. They must issue notices to more people than they expect to layoff because of notification timelines in contracts, and they must ensure that they do not leave positions open so that others can "bump" or move into them because of seniority. Thus, an individual in one area who possesses minimal qualifications but has seniority could, after receiving layoff notice, "bump" or replace someone in another area. Such a condition ignores expertise in the labor force and undermines the quality of service for students and teaching. It turns the community college into an economic and political arm of the state and trivializes its educational function. Students, especially the most disadvantaged, are victims of the state's political capriciousness.

The most disadvantaged college students are the least likely to benefit from a political economy that is steered by a neoliberal ideology. Former Secretary of Labor in the Clinton administration, Robert Reich, envisioned the graduates of community colleges sustaining a new high-technology economy, with "high wages" and "high skills."[11] Other observers assumed that the growth of the U.S. economy would be fueled by students-as-products of community colleges.[12] However, large numbers of those students who attend community colleges are not headed for personal economic prosperity or for critical places in the expanding economy. Instead, they are engaged in survival and looking for a slot in the service economy. These students, largely ignored by scholars and observers of higher education, include the immigrants—documented and undocumented—who undertake English as a second language programs; African American and Hispanic/Latino welfare clients who enroll in noncredit medical terminology courses so that they can qualify for entry-level medical office jobs; and physically and mentally disabled students who endeavor to learn practical techniques to get through the week or day—such as navigating the local bus schedule or completing a job application. I refer to this population as "students beyond the margins" to distinguish them from predominantly mainstream credit students or from students who are identified with "marginalized" groups, typically those with minority status, but who have no other nontraditional characteristics. To what extent are community colleges—whose mission includes service to the underserved—accommodating and educating "students beyond the margins"?

With an historical commitment to access and to a comprehensive curriculum, and with a mandate—often legislated—for economic development and job preparation, the community college is conflicted about its obligation to students, particularly highly nontraditional students or disadvantaged students. This conflict is especially evident during periods when economic issues are institutional priorities.[13] When fiscal retrenchment is required, programs and services for mainstream students receive priority, with cuts going to programs for those students who are beyond the margins. Nonetheless, individual administrators and faculty, who are proponents of access for students beyond the margins, act to support these students and their programs even to the extent that they ignore or violate policies and past practices—all in order to provide justice to students.

For the compensatory education students at Johnston Community College in North Carolina, their program and their progress are dependent upon those college employees—such as the president and the instructor—who support and accommodate them. Specifically, it is the instructor who teaches them how to read and how to survive in their communities. It is thus under these circumstances and at this place where these students experience justice as they are treated fairly and with dignity. In this environment, they are advantaged.

For the adult education students at Pima Community College, there is the adult basic education administrator who champions their cause, meets with legislators to keep the program funded, and manages a program that provides basic educational skills for students who have been abandoned. The students are not only disadvantaged educationally but also disenfranchised politically. Justice for them is in the hands of Pima College, and specifically the ethical responsibility of this dean.

For the welfare students at Edmonds Community College, their only economic hope is the upgrading of their skills so that they can find employment. It is the potential promise for their children that motivates them. In spite of the demands and pressures of their lives and the program requirements, these students do finish their program, graduate, and improve their economic station. The gains may be marginal, but they are significant. The program director and the program faculty act in tandem to rescue these students and provide not just education and training, but life opportunities as well.

For students without much in the way of prospects for postsecondary education, a program administrator at Bakersfield College finds the money to support their education and training and then motivates the students to pursue further education. He recognized that without direction these nontraditional students would not advance. He exemplifies those faculty and administrators who privilege not those with advantage but those who are disadvantaged, conforming to John Rawls' characterization of justice.[14]

Notwithstanding the numerous examples of community colleges advantaging the disadvantaged, in the main, individuals, not groups, are accommodated and even provided equitable outcomes through one of two mechanisms. Students are advantaged first through specific programs, such as work-study, tutoring, "first generation," financial aid that includes tuition remission, and academic cohort programs; and second, through the efforts of individual administrators and faculty, whom I designate autonomous agents—those who champion the cause of nontraditional students and individual students. Highly nontraditional and "beyond the margins" students have four specific categories of needs. First, they need guides and mentors. Second, they need financial aid, which includes tuition remission, grants, and paid employment. Third, they need a peer community. And fourth, they need academic integration that includes an academic plan that will move them toward specific goals.[15] These needs are met on an individual basis for relatively small populations of students; this is not a systematic approach taken by institutions or by state governments.

Although higher education as an institution may not have fulfilled its obligation "to treat all persons equally, to provide genuine equality of opportunity . . . [by giving] more attention to those with fewer native assets and to those born into the less favorable social positions,"[16] there are local examples where justice is accorded to students. Within the community college, thousands of students are beyond the margins; more than "at risk," they are desperate and they come to the community college as a final refuge. With difficulty, some of these students at community colleges are accommodated and justly served by programs run by administrators, faculty, and staff who are neither arms of the state nor servants of institutional bureaucracy. Thus, justice as fairness is enacted at these community colleges through the work of faculty and administrators, whose focus is upon those they find in need of rescue from their social, economic, and physical limitations. These disadvantaged students are brought in for shelter and rejuvenation.

Much of the actions of reform are aimed at rectifying the outcomes of state policies as well as the funding behaviors of state and federal governments. Although the community college is arguably an extension of a capitalistic regime formalized in the state,[17] community college actors work to remediate the ills of capitalism. In the United States, the policies for a globally competitive workforce have engendered an educational system that privileges academic performativity in the form of skills. At the same time, it punishes "laggards" who cannot keep up. This condition of "winners and losers" is a replay of geopolitics as poor countries suffer as wealthy countries prosper.[18] This is what Brecher, Childs, and Cutler refer to as "globalization from above," where power and resources move toward elites and away from communities.[19]

A number of behaviors within U.S. community colleges in the past decade have exhibited this movement toward elites and away from communities.[20] Yet, this pattern is not universal either from institution to institution or within institutions. While this examination identifies autonomous agents[21] as resisters of this pattern and actors for justice, there are other factors and conditions that enable action directed to justice for disadvantaged students. These factors and conditions include the organization, management, and governance of individual colleges to the extent that either considerable autonomy is accorded organizational members or a high level of organizational slack exists, as typically found in human service organizations and higher education institutions.[22] Under these conditions, actors are not monitored closely. Or, as in the case of the president of Johnston Community College in North Carolina, the institutional leader, with the authority and power of the office as well as a personal conviction to aid the needy, can engender actions for justice within the institution. Such individuals in positions of authority and autonomy have considerable power.

In community colleges, particularly in specific program areas, faculty and administrators act to grant justice to their students. This pattern of behavior is consistent with Rawls: Educational institutions have a responsibility to ensure substantive equality of opportunity. Disadvantaged students must not be subjected to an educational system or program in which their individual agency and self-purpose are neglected in favor of the economic benefit for local industry. National or local economic competitiveness cannot justify the commodification of students, in which their rights to equality of opportunity are sacrificed for a larger, rationalized good. Institutional members have the capacity to enact their understandings of the community college mission, and that mission includes accommodating and advantaging students, especially disadvantaged students.

The Problem of Disadvantaged Students

Disadvantaged students are those who by birth or circumstances (such as illness, plant closure, domestic strife, and the like) have barriers that either prevent or constrain their acquisition of basic rights and duties.[23] The purpose of educational institutions in a just society is thus to improve the position of these students in the acquisition of these rights and duties. Improvement signifies that institutions have advantaged students. The problem, however, is that increasingly our higher education institutions are adopting neoliberal policies that favor privileged populations.[24]

Conditions in U.S. society within the new economy favor those who have social and economic advantages. While those in positions of privilege are further privileged—for example, they gain more wealth—those who are economically distressed continue to face more severe hardships, including poverty. Personal benefits are slow to come to those who have been disadvantaged within their lifetime and to those connected to groups disadvantaged historically over generations. The flourishing of individuals, while certainly part of U.S. social and cultural conceptions of the good, is more likely for some populations than for others. Although access to higher education and to the acquisition of its benefits, such as economic and social mobility, appears to be part of both the distribution and gaining of advantage, specific student populations do not receive and gain as much as others through this advantaging. The difference between access to a remedial program and to a university transfer program can be considerable, as can the difference between access to a community college and to a select liberal arts college. Within these categories, the difference between the completion of a university transfer program with an accumulated grade point average of 2.0 (or C) and completion with a 3.5 (or B+) grade point average is also significant.[25] Is the difference based upon social and economic inequalities? Or were these inequalities addressed by the institution during a student's educational career? Was there fair and reasonable distribution of resources to advantage disadvantaged students?

These answers define the very condition of our society and perhaps its future. Our institutions and state systems of higher education are discriminatory in their treatment of students. Students with disabilities, students who are in programs that have low institutional prestige, and students who are not legal immigrants to the United States may be and are treated different from other students. Government policies are also discriminatory: The welfare population of single parents has constraints upon them as students that are not imposed upon other students; and large numbers of part-time students are not qualified for financial aid. In denying what I have referred to as "advantaging disadvantaged students"—indeed, in disadvantaging some student populations further—institutions of higher education and governments are failing to advance our society to a condition where there is indeed justice.

Solutions: Justice Is the First Imperative of Our Institutions

Neoliberal ideology and its related practices are antithetical to justice for disadvantaged populations. Although neoliberal policies and practices have found a home in higher education institutions, there is considerable evidence that these institutions are neither appropriate sites for neoliberal practices nor compatible in their mission and purposes with the ideology.[26] For the community college, the economic and competitive orientation over the past twenty-five years has skewed the access mission and compromised quality by treating students as economic commodities while the institution increasingly served markets in favor of communities.[27] The public good as well as individual student development in effect became subordinate to commercial and economic concerns—raising money, training for the private sector, and improving institutional productivity and efficiency. As well, government priorities, such as global competitiveness (e.g., free trade), national security, energy, tax reduction, welfare reduction, and a general decrease in the role of the state in the public domain, helped to reframe the community college as an extension of the state and, to some extent, as a modified social safety net. Thus, as a consequence of job loss in manufacturing sectors due to the globalization of production, community colleges became sites of retraining laid-off workforces. Such retraining was primarily at the lower end of educational skills, with minimal support services such as personal and career counseling available, and little or no promise of similar jobs after college. The community college was equally and poorly used as a site for the education of welfare recipients.[28] In these cases of worker displacement and welfare reform, work, not education, was the state's priority, and the work envisaged was at low levels of skill and remuneration. Whereas, historically, the community college has opened its doors to the underserved of society, the state was, true to neoliberal ideology, closing its doors. Reluctantly perhaps, the community college facilitated the movement of the state away from responsibility for the welfare of its citizens. For those who are not citizens, the privileges are meager. Undocumented immigrants are indeed treated as a national problem.[29] The state's position on these immigrants—at the federal level and, with some variation, at the state level—is ambivalent. On the one hand, this population serves the economy as a workforce. On the other hand, undocumented immigrants lack entitlement to public services at a level equal to that of others. As students in higher education, they are ineligible for federal financial aid.[30] And while each state can determine who shall pay in-state tuition and who must pay out-of-state tuition, states can also deny access to undocumented immigrants. Both actions—the denial of access and the requirement of out-of-state tuition—are punishments for an already disadvantaged population.

There are, nonetheless, remarkable actions at institutions advantaging the disadvantaged. Among these are the Middle College at GateWay Community College in Arizona; the case management approach to students at Community College of Denver; first-generation programs at Johnston Community College in North Carolina; the Science, Technology, Engineering, and Mathematics (STEM) program for first-generation and minority students at Bakersfield College in California; and work-study programs at Edmonds Community College in Washington. As well, there are countless instances of individual faculty, staff members, and administrators who assist students to advantage them—at Virginia Highlands Community College in Virginia, at Borough of Manhattan Community College in New York, at Mountain View College in Texas, and at El Camino College in California. Behaviors include waiving tuition fees for economically distressed students, giving individual attention to troubled students, mentoring students, as well as bending or even breaking the rules for the benefit of students. Additionally, at the policy level, chief executives make it their life's work to better the lives of students, particularly disadvantaged students. We could conclude that surely these actions should be enough to enact justice.

Many of the behaviors and actions, however, are responses or reactions to prevailing conditions. They constitute triage, not preventive action. The prevailing conditions include not only students' economic conditions, but also institutional paucity of resources, punitive federal and state policies, and civic neglect.[31] Although there are numerous models and examples of practices that aid disadvantaged students, they have limited impact compared to the weight of the problem. Yet, there are several directions that legislators, policy influencers, and institutional officials could follow.

Financial Support from Government

Community colleges require state and federal financial support that is equitable. By this I mean that program costs need to be factored into allocations, and these costs should be relatively equal. Thus, the cost of an instructor for an electronics course should be approximately equal to the cost of an instructor for an English as a second language course. At present, funding in some states for basic and compensatory education is at a rate less than other program areas, and the use of part-time, lesser paid instructors is more common in lesser status courses and programs.[32] Cost differentials for programs would apply in such programs as nursing, where instructor and student ratios are less than in liberal arts—that is, the actual costs in nursing are greater, not because of employee compensation, but because of pedagogical or legal issues (e.g., in nursing and laboratory courses, there are legal maximums of students that instructors are permitted to supervise). In this way, all programs would receive fair treatment and students in some programs would not be unduly privileged because they are in elite programs.

Public and Private Uses of Educational Funding

While greater levels of funding are required for community colleges to sustain their activities and to meet future predicted growth in student numbers, as well as to move these colleges closer to parity with other postsecondary educational institutions, community colleges should not use their funding from public sources to serve strictly private interests.[33] Although David Labaree points out that there is both historical tension and struggle between democratic and capitalistic goals of community colleges,[34] institutional actions that favor market-based goals have prevailed over the past two decades.[35] In order to restore at least a balance between the democratic and capitalistic goals, public monies must be put to more public uses, such as paying attention to individual student development and mobility and to the social good. Part of this entails granting benefits to the disadvantaged so that they can achieve upward economic and social status and so that their futures will be gains, not burdens, for society.

Universal Access

Programs and services for students must be universal: They must be open to the entire population of any targeted group, such as first-generation students, single parents, low economic status students, and disabled students. At present, colleges do not have sufficient resources to meet this standard, and thus their programs are restricted. Numerous students and college officials who responded to my questions expressed considerable satisfaction with institutional programs for special populations. Indeed, a number of students in a recent follow-up investigation indicated that without such programs as first generation, they would not persist in college.[36] Others who were supported financially by the institution claimed they would not have attended college without this assistance. But the majority of students who were in need of programs and assistance did not benefit because the institution's resources were not adequate to provide universal access. As Norton Grubb has noted in assessing California's programs for disconnected youth, there are numerous small programs, but they do not serve an entire population and not all are effective.[37] The prescription is to have fewer but larger programs that serve greater populations more effectively.

Public Education

The magnitude of the problem of providing education and training to a broad, diverse, and often educationally underprepared population of students at community colleges is neither evident to the public nor apparent to policymakers and legislators.[38] The public continues to think of college students as traditional—recent high school graduates who go to college full-time. Institutions, agencies, and foundations can and should endeavor to provide support for public education on the community college. All too often, community colleges and their supporters devote their energies and resources to promoting the institution and advertising its achievements and positive characteristics. There are no stories about students who are badly served by the institution and society: individuals whose conditions keep them from attending college, or students who drop out with poor grades. Large numbers of community colleges students are on the periphery of social and economic life; there are others whose disabilities—mental or physical—keep them at low levels of attainment. Their conditions are not conveyed publicly.

Justice as Fairness

There are countless numbers of students in community colleges who have less than privileged lives or advantaged circumstances. Doug lost his job, lived in his car for six months, and suffered several accidents that led to sepsis in his leg and near amputation and death. He worked for years as a truck driver, but his health is not good enough for him to continue. As well, he wants a more mentally challenging and secure job. He returns to college at age 40 to try to attain a career in water resource management, and although he has a painful life story, he expresses a positive view about learning. Michelle is a former alcoholic, drug addict, prostitute, and biker in her late thirties who is studying and working her way through community college to move on to a social work degree. Patricia, a fifty-three-year-old African American who is disabled and for years was institutionalized by her family for psychiatric problems, has made her way from community outreach programs to college preparation courses, to a certificate program in accounting. She is on route to a degree in accounting, and her grades are strong. These students have moved a considerable distance from their social, personal, and economic conditions to attend college, and through college they are endeavoring to gain advantages so that they can be the same as or equal to others. There is little doubt that these three individuals have been rescued by the community college: Doug from despair; Michelle from death; and Patricia from loneliness and a life without a vocation. There are others who do not fare as well. Doug, Michelle, and Patricia were aided by college programs for special populations and by instructors who devoted more than was required to them. Not every student is this fortunate.

To reiterate the points offered in Chapter 2, and applying the principles of justice from John Rawls, institutions must be rated by how effectively they guarantee the conditions necessary for all students, equally, to further their aims, or by how efficiently they advance shared ends that will similarly benefit everyone. In other words, one can judge a state's educational system by how well it facilitates real, not merely formal, equal opportunity for the worst-off citizen.[39] One can easily assume that developmentally challenged students, for example, are among the worst off in society, as they lack those basic skills required for independent living.

According to Rawls, educational institutions have a responsibility to ensure substantive equality of opportunity, regardless of the potential economic benefits of unequal access. Disadvantaged students must not be subjected to an educational system or degree program in which their individual agency and self-purpose are neglected in favor of the economic benefit for a local industry. According to Rawls, national, or indeed local, economic competitiveness cannot justify the commodification of students, in which their rights to equality of opportunity are sacrificed for a larger good. This is the case even if greater competitiveness would increase the net utility in society. That larger good becomes irrelevant once it is demonstrated that (a) the student does not enjoy substantive equality of opportunity in an educational system and (b) the disadvantaged student is not better off as a result of the political-economic distributive scheme.

Thus, while students such as Doug, Michelle, and Patricia are advantaged by their community college education and experiences, this does not mean—and the evidence suggests this is not the case—that all disadvantaged students are advantaged by the community college. The "success" stories are not universals but merely the minority or even exceptional cases. Furthermore, for those disadvantaged populations that either cannot or do not access the community college—for example, the millions referred to by the chancellor of the Community Colleges of California as "people who are simply typically lower SES kids . . . who never find the first rung on the social ladder . . . probably 1.2 million [in California]"—they certainly do not benefit from the community college, from either its learning programs or its promises for economic and social mobility. As an arm of the state, the community college cannot simply remain an economic instrument but must also work toward the provision of justice for those who could benefit from college education.

The Refrain

Conceptions of nontraditional students have been narrow, possibly based upon a limited population and focused primarily upon institutional interests, such as institutional goals and the need for resources. These conceptions and the associated practices of institutions, including colleges and the state, have 1) ignored some student needs, 2) led to diminished responses to some needs, 3) addressed a small portion of the population, and 4) elevated a category of student behaviors (e.g., program completion, university transfer, and job attainment) beyond other categories (e.g., skill and personal development, social development and citizenship, and self-worth). That is, institutional goals have taken precedence over students' goals and needs. Notwithstanding these outcomes, community colleges have both sustained their rhetorical commitment to undeserved populations and responded within the bounds of their limited resources to these populations. Indeed, there are large numbers of faculty, staff, and administrators at colleges who act deliberately to minister to the needs of disadvantaged students. These actions are responses to conditions perpetuated by exogenous forces—by federal and state policy, by family economic status and domestic conditions, by previous school experiences, and by employers and agencies. One example is financial discrimination against undocumented immigrants who are required to pay out-of-state tuition for college, which can be as much as five times that of in-state tuition, leading to either severe financial hardships or simply nonattendance. Exogenous forces also include abusive relationships in which children or adult women are removed or remove themselves from the home and face economic hardships. They also include high school experiences that result in noncompletion of a course of study that would both credential students and supply them with academic skills to cope with either the labor market or further education. Finally, these forces include state and federal governments' unwillingness to provide financial support to colleges so that costs are not passed on to students—costs that either constrain students' educational behaviors (e.g., forcing them to work instead of take classes and study) or deny students access to an education.

To some extent, what is at issue here is ideology—the belief of the state and its citizens in justice or fairness and the means to achieve them. For neoliberals and neoconservatives, social welfare and its attendant behaviors are unacceptable; instead, private, individual action is the only remedy for disadvantaged populations. Individual responsibility and private sector largesse are viewed as keys to societal development and, ultimately, economic prosperity, as if these were justifiable substitutes for justice.

In Rawls' sense of justice, the common or social good takes precedence over individual prosperity, and a social contract among citizens is understood as the basis of a well-ordered and just society. The means to achieve justice—whether in a welfare state or a capitalistic postindustrial state—is secondary as long as those individuals whose conditions, defined by birth or life's experiences, place them at a disadvantage are improved or advantaged.

Students in higher education institutions are not equal. Some are more privileged than others; some are wealthier, more able, and better prepared than others. The ethical responsibility of our institutions is not to make all students equal but rather to treat students fairly so that those who are less privileged and less advantaged than others are accorded justice: that is, given advantages and benefits so that their disadvantages are nullified.

Thus, the ways in which we view and assess or judge higher education institutions must alter so that justice is a preferred outcome. Similar to Alexander Astin's concept of the development of talent as a valid measure of an institution's worth,[40] the practice of fairness can be measured by the advantaging of the disadvantaged. The worth of an institution depends upon what it has added or altered for its participants and constituents. For the marginally literate to advance through adult high school at a community college is not only an accomplishment but also an act of justice and an achievement for a college.

Notes

1. Steven Brint and Jerome Karabel, *The Diverted Dream: Community Colleges and the Promise of Educational Opportunity in America, 1900–1985* (New York: Oxford University Press, 1989); Deborah L. Floyd, Michael L. Skolnik, and Kenneth P. Walker, eds., *The Community College Baccalaureate: Emerging Trends and Policy Issues* (Sterling, VA: Stylus, 2004); John H. Frye, *The Vision of the Public Junior College, 1900–1940* (New York: Greenwood, 1992); W. Norton Grubb and Marvin Lazerson, *The Education Gospel* (Cambridge, MA: Harvard University Press, 2004); Ken Meier, "The Community College Mission: History and Theory" (unpublished manuscript, Bakersfield, CA: 2004); James Ratcliff, "Seven Streams in the Historical Development of the Modern Community College," in *A Handbook on the Community College in America*, ed. G. Baker, 3–16 (Westport, CT: Greenwood, 1994).

2. Scott Davies and Neil Guppy, "Globalization and Educational Reforms in Anglo-American Democracies," *Comparative Education Review* 41, no. 4 (1997): 435–59; Rosemary Deem, "'New Managerialism' and Higher Education: The Management of Performances and Cultures in Universities in the United Kingdom," *International Studies in Sociology of Education* 8, no. 1 (1998): 47–70.

3. Arthur Cohen and Florence Brawer, *The American Community College* (San Francisco: Jossey-Bass, 2003); John S. Levin, "Neoliberal Policies and Community College Faculty Work," in *Handbook of Higher Education*, vol. 22, ed. William Tierney, (Norwell, MA: Kluwer Academic Publishers, forthcoming).

4. John S. Levin, "The Business Culture of the Community College: Students as Consumers; Students as Commodities," in "Arenas of Entrepreneurship: Where Nonprofit and For-profit Institutions Compete," ed. Bruce Pusser, special issue, *New Directions for Higher Education* 129 (2005): 11–26.

5. Kent A. Phillippe and Leila Gonzalez Sullivan, *National Profile of Community Colleges: Trends and Statistics*, 4th ed. (Washington, DC: American Association of Community Colleges, 2005).

6. Marcia Baxter Magolda, *Making Their Own Way: Narrative for Transforming Higher Education to Promote Self-Development* (Sterling, VA: Stylus, 2001); Dorothy C. Holland and Margaret A. Eisenhart, *Educated in Romance: Women, Achievement and College Culture* (Chicago: University of Chicago Press, 1990); Michael Moffat, *Coming of Age in New Jersey: College and American Culture* (New Brunswick, NJ: Rutgers University Press, 1989); Rebekah Nathan, *My Freshman Year: What a Professor Learned by Becoming a Student* (Ithaca, NY: Cornell University Press, 2005).

7. Tom Wolfe, *I Am Charlotte Simmons* (New York: Farrar, Straus and Giroux, 2004).

8. Marion Bowl, *Nontraditional Entrants to Higher Education* (Stoke on Trent, UK: Trentham Books, 2003); Tim Silva et al., "Adult Education Participation Decisions

and Barriers: Review of Conceptual Frameworks and Empirical Studies" (prepared for U.S. Department of Education, Office of Educational Research and Development, National Center for Education Statistics, August 1998).

9. Cohen and Brawer, *The American Community College*.

10. On March 12 and 13, 2003, Bakersfield College had to issue layoff notices, although few of these notices resulted in employment termination at the college that year. A budget shortfall—created by a reduction in state appropriations in both 2003 and 2004—led to college plans to reduce the labor force.

11. Robert Reich, "Hire Education," *Rolling Stone* (October 20, 1994): 119–25; Robert Reich, *The Work of Nations: Preparing Ourselves for Twenty-First Century Capitalism* (New York: Vintage Books, 1992).

12. Anthony P. Carnavele and Donna M. Desrochers, "Community Colleges in the New Economy," *Community College Journal* 67, no. 5 (April/May 1997): 26–33; Mary Ann Roe, *Education and U.S. Competitiveness: The Community College Role* (Austin: IC2 Institute, University of Texas, 1989).

13. Levin, *Globalizing the Community College: Strategies for Change in the Twenty-First Century* (New York: Palgrave, 2001).

14. John Rawls, *A Theory of Justice* (Cambridge, MA: Belknap Press of Harvard University Press, 1999).

15. Levin, "Nontraditional Students and Community Colleges: The Conflict of Justice and Neoliberalism. An Overview" (paper presented at the American Educational Research Association, Montreal, April 2005).

16. Rawls, *A Theory of Justice*, 86.

17. Martin Carnoy, *The State and Political Thought* (Princeton, NJ: Princeton University Press, 1984).

18. Gary Teeple, *Globalization and the Rise of Social Reform* (Atlantic Highlands, NJ: Humanities Press, 1995).

19. Jeremy Brecher, John Brown Childs, and Jill Cutler, *Global Visions: Beyond the New World Order* (Boston: South End, 1993).

20. Levin, *Globalizing the Community College*.

21. Martin Lipsky, *Street-Level Bureaucracy* (New York: Russell Sage Foundation, 1980).

22. Yeheskel Hasenfeld, *Human Service Organizations* (Englewood Cliffs, NJ: Prentice Hall, 1983); James March and Michael Cohen, *Leadership and Ambiguity: The American College President* (New York: McGraw-Hill, 1974); John Meyer and Brian Rowan, "Institutionalized Organizations: Formal Structure as Myth and Ceremony," *American Journal of Sociology* 83 (1977): 340–63; Karl Weick, "Educational Organizations as Loosely Coupled Systems," *Administrative Science Quarterly* 21 (1976): 1–19.

23. John Rawls, *Political Liberalism* (New York: Columbia University Press, 1993).

24. Sheila Slaughter and Gary Rhoades, "The Neoliberal University," *New Labor Forum* (Spring/Summer 2000): 73–79.

25. The remarkable differences between students who make a passing grade in a course and those who receive high grades lead to considerable differences in educational attainment, an observation made by Estela Bensimon in her presentations on "The Diversity Scorecard." See, for example, "Equity as a Fact and Equity as a Result," presidential invited session, (American Educational Research Association, Montreal, April 13, 2005).

26. John Levin, Susan Kater, and Richard Wagoner, *Community College Faculty: At Work in the New Economy* (New York: Palgrave Macmillan, 2006); Levin, "The

Business Culture of the Community College: Students as Consumers; Students as Commodities," in *Arenas of Entrepreneurship: Where Nonprofit and For-Profit Institutions Compete*, vol. 129 of *New Directions for Higher Education*, ed. B. Pusser, 11–26 (San Francisco: Jossey-Bass); Slaughter and Rhoades, *Academic Capitalism and the New Economy: Markets, State, and Higher Education* (Baltimore: Johns Hopkins University Press, 2004); Slaughter and Rhoades, "The Neoliberal University."

27. Alicia C. Dowd, "From Access to Outcome Equity: Revitalizing the Democratic Mission of the Community College," in "Community Colleges: New Environments, New Directions," ed. Kathleen Shaw and Jerry Jacobs, special issue, *The Annals of the American Academy of Political and Social Science* 586, no. 1 (2003): 92–119; Levin, *Globalizing the Community College*; Levin, "Business Culture"; John Roueche and George A. Baker III, *Access and Excellence* (Washington, DC: Community College Press, 1987); George Vaughan, "The Big Squeeze at Community Colleges," *The News & Observer* (March 24, 2002).

28. Jerry A. Jacobs and Sarah Winslow, "Welfare Reform and Enrollment in Postsecondary Education," *The Annals of the American Academy of Political and Social Science* (March 2003): 194–217; Christopher Mazzeo, Sara Rab, and Susan Eachus, "Work-First or Work-Study: Welfare Reform, State Policy, and Access to Postsecondary Education," *The Annals of the American Academy of Political and Social Science* (March 2003): 144–71; Shaw et al., *Putting Poor People to Work: How the Work-First Ideology Eroded College Access for the Poor* (unpublished manuscript, 2005); Kathleen Shaw and Sara Rab, "Market Rhetoric versus Reality in Policy and Practice: The Workforce Investment Act and Access to Community College Education and Training," *The Annals of the American Academy of Political and Social Science* (March 2003).

29. Radha Roy Biswas, *Access to Community College for Undocumented Immigrants: A Guide for State Policymakers* (Boston: Lumina Foundation for Education and Jobs for the Future, 2005).

30. Katalin Szelényi and June C. Chang, "Educating Immigrants: The Community College Role," *Community College Review* 30, no. 2 (2002): 55–73.

31. In interviews with civic and state public officials in 2006, I note that these individuals are not well informed about the actual presence or plight of disadvantaged students at community colleges and are ill informed about the impact of state policy on these populations.

32. Levin, Kater, and Wagoner, *Community College Faculty*.

33. Levin, "Neoliberal Policies."

34. David Labaree, "Public Goods, Private Goods: The American Struggle over Educational Goals," *American Educational Research* Journal 34, no. 1 (1997): 39–81.

35. Levin, *New World College: Community Colleges in the New Economy* (Toronto, ON: 2006).

36. John S. Levin and Jeremy Levin, *The Costs of College*, DVD, Riverside, CA: 2006.

37. Grubb, "Using Community Colleges to Reconnect Disconnected Youth" (Menlo Park, CA: William and Flora Hewlett Foundation, 2003).

38. This lack of awareness was brought to my attention during interviews with college and system executives during the period of 2003–6, and civic and state political officials in 2006.

39. Rawls, *A Theory of Justice*.

40. Alexander W. Astin, *Achieving Educational Excellence* (San Francisco: Jossey-Bass, 1985).

Bibliography

Academic Senate of California's Community Colleges. *Scenarios to Illustrate Effective Participation in District and College Governance*. Community College League of California and the Academic Senate for California Community Colleges, 1996. Available at http://www.academicsenate.cc.ca.us/Publications/Papers/FinalScenario .htm. Accessed June 23, 2005.

Adelman, Clifford. *Moving into Town And Moving On: The Community College in the Lives of Traditional Age Students*. Washington, DC: U.S. Department of Education, 2005.

Alfred, Richard, and Patricia Carter. "New Colleges for a New Century: Organizational Change and Development in Community Colleges." In *Higher Education: Handbook of Theory and Research*, edited by John C. Smart and William G. Tierney, 240–83. New York: Agathon, 1999.

Allen, Betty A. "The Student in Higher Education: Nontraditional Student Retention." *Community Services CATALYST* 23, no. 3 (1993): 19–22.

American Association of Community Colleges. "Our Mission Statement." American Association of Community Colleges, January 2005. Available at http://www.aacc.nche .edu/Content/NavigationMenu/AboutAACC/Mission/OurMissionStatement.htm.

American Association of State Colleges and Universities. "Policy Matters: Should Undocumented Immigrants Have Access to In-State Tuition?" *AASCU newsletter* 2, no. 6 (2005).

Anderson, Benedict. *Imagined Communities: Reflections on the Origin and Spread of Nationalism*. New York: Verso, 1991.

Appadurai, Arjun. "Disjunctures and Difference in the Global Cultural Economy." In *Global Culture: Nationalism, Globalization, and Modernity*, edited by Mike Featherstone, 295–310. Newbury Park, CA: Sage, 1990.

Apple, Michael. "Comparing Neoliberal Projects and Inequality in Education." *Comparative Education* 37, no. 4 (2001): 409–23.

Arnason, Johann. "Nationalism, Globalization, and Modernity." In *Global Culture: Nationalism, Globalization, and Modernity*, edited by Mile Featherstone, 207–36. Newbury Park, CA: Sage, 1990.

Aronowitz, Stanley, and William Di Fazio. *The Jobless Future: Sci-tech and the Dogma of Work*. Minneapolis: University of Minnesota Press, 1994.

Aslanian, Carol. "You're Never Too Old: Excerpts from Adult Students Today." *Community College Journal* 71, no. 5 (2001): 56–58.

Astin, Alexander W. *Achieving Educational Excellence*. San Francisco: Jossey-Bass, 1985.

Astin, Alexander, and Leticia Oseguera. "The Declining 'Equity' of American Higher Education." *The Review of Higher Education* 27, no. 3 (2004): 321–41.

Ayers, D. Franklin. "Discursive Manifestations of Neoliberal Ideology in Community College Mission Statements: A Critical Discourse Analysis." Unpublished manuscript, University of North Carolina at Greensboro, 2004.

———. "Neoliberal Ideology in Community College Mission Statements: A Critical Discourse Analysis." *The Review of Higher Education* 28, no. 4 (2005): 527–49.

Bagnall, Richard. "Lifelong Learning and the Limitations of Economic Determinism." *International Journal of Lifelong Education* 19, no. 1 (2000): 20–35.

Bailey, Thomas R., and Mariana Alfonso. *Paths to Persistence: An Analysis of Research on Program Effectiveness at Community Colleges*. New York: Teachers College, Columbia University, 2005.

Bailey, Thomas R., Mariana Alfonso, Juan Carlos Calcagno, Davis Jenkins, Gregory S. Kienzl, and Timothy Leinbach. *Improving Student Attainment in Community Colleges: Institutional Characteristics and Policies*. New York: Community College Research Center, Teachers College, Columbia University, 2004.

Bailey, Thomas R., Mariana Alfonso, and Marc Scott. *The Education Outcomes of Occupational Postsecondary Students*. New York: Community College Research Center, Teachers College, Columbia University, 2005.

Bailey, Thomas R., and Irina E. Averianova. "Multiple Missions of Community Colleges: Conflicting or Complementary." Occasional paper, Community College Research Center, Teachers College, New York, 1998.

Bailey, Thomas R., Juan Carlos Calcagno, Davis Jenkins, Gregory S. Kienzl, and Timothy Leinbach. *The Effects of Institutional Factors on the Success of Community College Students*. New York: Community College Research Center, Teachers College, Columbia University, 2005.

Bailey, Thomas R., Davis Jenkins, and Timothy Leinbach. "Is Student Success Labeled Institutional Failure? Student Goals and Graduation Rates in the Accountability Debate at Community Colleges." New York: Community College Research Center, Teachers College, Columbia University, 2005.

Bailey, Thomas R., and Vanessa Smith Morest. *The Organizational Efficiency of Multiple Missions for Community Colleges*. New York: Teachers College, Columbia University, 2004.

Baker, Carolyn. "Ethnomethodological Analyses of Interview." In *Handbook of Interview Research: Context and Method*, edited by Jaber Gubrium and James Holstein, 777–95. Thousand Oaks, CA: Sage, 2002.

Bankston, Carl. "Immigrants in the New South: An introduction." *Sociological Spectrum* 23, no. 2 (2003): 123–28.

Baxter Magolda, Marcia. *Making Their Own Way: Narrative for Transforming Higher Education to Promote Self-Development*. Sterling, VA: Stylus, 2001.

Bay, Libby. "Twists, Turns, and Returns: Returning Adult Students." *Teaching English in the Two-Year College* vol. 26, no. 3 (1999): 305–12.

Berker, Ali, Laura Horn, and C. Dennis Carroll. *Work First, Study Second: Adult Undergraduates Who Combine Employment and Postsecondary Enrollment*. Washington, DC: National Center for Education Statistics, 2003.

Birnbaum, Robert. *Management Fads in Higher Education: Where They Come From, What They Do, Why They Fail*. San Francisco: Jossey-Bass, 2000.

Biswas, Radha R. *Access to Community College for Undocumented Immigrants: A Guide for State Policymakers*. Boston: Lumina Foundation for Education and Jobs for the Future, 2005.

Bogart, Quentin. "The Community College Mission." In *A Handbook on the Community College in America*, edited by George Baker, 60–73. Westport, CT: Greenwood, 1994.

Bok, Derek. *Universities in the Marketplace: The Commercialization of Higher Education*. Princeton, NJ: Princeton University Press, 1995.

Bowen, William, and Derek Bok. *The Shape of the River: Long-Term Consequences of Considering Race in College and University Admissions*. Princeton, NJ: Princeton University Press, 1998.

Bowl, Marion. *Nontraditional Entrants to Higher Education*. Stoke on Trent, UK: Trentham Books, 2003.

Bowland, Gay. "A Fresh Start." *Community College Week* (August 16, 2004): 6–8.

Brecher, Jeremy, John Brown Childs, and Jill Cutler. *Global Visions: Beyond the New World Order*. Boston: South End, 1993.

Brint, Steven. "Few Remaining Dreams: Community Colleges since 1985." *The ANNALS of the American Academy of Political and Social Sciences* (March 2003): 16–37.

Brint, Steven, and Jerome Karabel. *The Diverted Dream: Community Colleges and the Promise of Educational Opportunity in America, 1900–1985*. New York: Oxford University Press, 1989.

Burgess, Robert. *In the Field: An Introduction to Field Research*. London: George Allen and Unwin, 1984.

———. *Strategies of Educational Research: Qualitative Methods*. London: Falmer Press, 1985.

Caiazza, Amy, April Shaw, and Misha Werschkul. *Women's Economic Status in the States: Wide Disparities by Race, Ethnicity, and Region*. Washington, DC: Institute for Women's Policy Research, 2004.

California Community Colleges System, Strategic Plan Steering Committee. "California Community Colleges System Strategic Plan. Education and the Economy: Shaping California's Future Today," 2006.

Campbell, John, and Ove Pedersen. "Introduction: The Rise of Neoliberalism and Institutional Analysis." In *The Rise of Neoliberalism and Institutional Analysis*, edited by John Campbell and Ove Pedersen, 2–23. Princeton, NJ: Princeton University Press, 2001.

Carnavele, Anthony P., and Donna M. Desrochers. "Community Colleges in the New Economy." *Community College Journal* 67, no. 5 (April/May 1997): 26–33.

———. *Help Wanted . . . Credentials Required: Community Colleges in the Knowledge Economy*. Washington, DC: Educational Testing Service and the American Association of Community Colleges, 2001.

———. "Why Learning? The Value of Higher Education to Society and the Individual." In *Keeping America's Promise: A Report on the Future of the Community College*, edited by K. Boswell and C. D. Wilson, 39–45. Denver: Education Commission of the States and the League for Innovation in the Community College, 2004.

Carnoy, Martin. *The State and Political Thought*. Princeton, NJ: Princeton University Press, 1984.

———. *Sustaining the New Economy: Work, Family, and Community in the Information Age*. Cambridge, MA: Harvard University Press, 2000.

Casey, Catherine. *Critical Analysis of Organizations: Theory, Practice, Revitalization*. Thousand Oaks, CA: Sage, 2002.

———. *Work, Society and Self: After Industrialism*. New York: Routledge, 1995.

Castells, Manuel. *The Rise of the Network Society*. 2nd ed. Malden, MA: Blackwell, 2000. First published 1996.

Chan Kopka, Teresita L., Nancy Borkow Schantz, and Roslyn Abrevaya Korb. *Adult Education in the 1990s: A Report on the 1991 National Household Education Survey*. Washington, DC: National Center for Education Statistics, 1998.

Chomsky, Noam. *Profit over People: Neoliberalism and Global Order*. New York: Seven Stories Press, 1999.

Choy, Susan P. "Nontraditional Undergraduates." In *The Condition of Education, 2002*, edited by U.S. Department of Education, 25–39. Washington, DC: U.S. Department of Education Office of Educational Research and Improvement, 2002.

Choy, Susan, and Larry Bobbitt. *Low-Income Students: Who They Are and How They Pay for Their Education*. Washington, DC: National Center for Education Statistics, 2000.

Chronicle of Higher Education. *Almanac Issue 2004–2005*. Washington, DC: Chronicle of Higher Education, 2004.

———. *College Enrollment by Age of Students, Fall 2003, The Chronicle of Higher Education, Almanac Issue 2005–2006*. Washington, DC: Chronicle of Higher Education, 2005.

———. "College Enrollment by Racial and Ethnic Group, Selected Years." *Almanac Issue, 2005–2006*. Washington, DC: Chronicle of Higher Education, 2005.

Clark, Burton. "The 'Cooling-Out' Function in Higher Education." *American Journal of Sociology* 65, no. 6 (1960): 569–76.

———. *Creating Entrepreneurial Universities: Organisational Pathways of Transformation*. Oxford: Pergamon, 1998.

———. *The Open Door College: A Case Study*. New York: McGraw-Hill, 1960.

Cohen, Arthur. "Governmental Policies Affecting Community Colleges: A Historical Perspective." In *Community Colleges: Policy in the Future Context*, edited by Susan Twombly, 3–22. Westport, CT: Ablex, 2001.

Cohen, Arthur, and Florence Brawer. *The American Community College*. San Francisco: Jossey-Bass, 2003. First published 1996.

Collins, Tom. *Contract Training as Revenue Source for Community Colleges*. Unpublished manuscript, Raleigh, NC, 2005.

Committee on Education and the Workforce. *Individuals with Disabilities Education Act (IDEA): A Guide To Frequently Asked Questions*. Available at http://edworkforce .house.gov/issues/109th/education/idea/ideafaq.pdf. Accessed in December 2005.

Community College Survey of Student Engagement. *Engaging Community Colleges: A First Look*. Austin, TX: Community College Leadership Program, 2002.

Creighton, Sean, and Lisa Hudson. *Participation Trends and Patterns in Adult Education: 1991 to 1999*. Washington, DC: National Center for Education Statistics, 2002.

Currie, Jan. Introduction to *Universities and Globalization*, edited by Jan Currie and Janice Newson, 1–13. Thousand Oaks, CA: Sage, 1998.

Currie, Jan, and Janice Newson, eds. *Universities and Globalization*. Thousand Oaks, CA: Sage, 1998.

Davies, Scott, and Neil Guppy. "Globalization and Educational Reforms in Anglo-American Democracies." *Comparative Education Review* 41, no. 4 (1997): 435–59.

De Angelis, Richard. "Globalization and Recent Higher Education Reforms in Australia and France: Different Constraints; Differing Choices in Higher Education Structure, Politics, and Policies." Paper for 9th World Congress on Comparative Education, Sydney, July, 1997.

Deem, Rosemary. "'New Managerialism' and Higher Education: The Management of Performances and Cultures in Universities in the United Kingdom." *International Studies in Sociology of Education* 8, no. 1 (1998): 47–70.

DeMartino, George. *Global Economy, Global Justice: Theoretical Objections and Policy Alternatives to Neoliberalism*. New York: Routledge, 2000.

Dennison, John, and Paul Gallagher. *Canada's Community Colleges*. Vancouver: University of British Columbia Press, 1986.

De Sousa, Jason. "Reexamining the Educational Pipeline for African-American Students." In *Retaining African Americans in Higher Education: Challenging Paradigms for Retaining Students, Faculty, and Administrators*, edited by Lee Jones, 21–44. Sterling, VA: Stylus, 2001.

DiMaggio, Paul, and William Powell. Introduction to *The New Institutionalism in Organizational Analysis*, edited by William Powell and Paul DiMaggio, 1–40. Chicago: University of Chicago Press, 1991.

Dougherty, Kevin. *The Contradictory College*. Albany: State University of New York Press, 1994.

Dougherty, Kevin, and Marianne Bakia. "Community Colleges and Contract Training: Content, Origins, and Impact." *Teachers College Record* 102, no. 1 (2000): 197–243.

———. "The New Economic Role of the Community College: Origins and Prospects." Occasional paper, Community College Research Center, Teachers College, New York, June, 1998.

Dowd, Alicia C. "From Access to Outcome Equity: Revitalizing the Democratic Mission of the Community College." In "Community Colleges: New Environments, New Directions." Edited by Kathleen Shaw and Jerry Jacobs. Special issue, *The ANNALS of the American Academy of Political and Social Science* 586, no. 1 (2003): 92–119

Dowd, Alicia C., and Randi Korn. "Students as Cultural Workers and the Measurement of Cultural Effort." Paper presented at the annual meeting of the Council for the Study of Community Colleges, Boston, April 8, 2005.

Dozier, Sandra Bygrave. "Undocumented and Documented International Students: A Comparative Study of Their Academic Performance." *Community College Review* 29, no. 2 (2001): 43–53.

Dudley, Janice. "Globalization and Education Policy in Australia." In *Universities and Globalization*, edited by Jan Currie and Janice Newson, 21–43. Thousand Oaks, CA: Sage, 1998.

Duggan, Lisa. *The Twilight of Equality? Neoliberalism, Cultural Politics, and the Attack on Democracy*. Boston: Beacon Press, 2003.

Edwards, Richard. "Lifelong Learning and a 'New Age' at Work." In *Lifelong Learning and Continuing Education: What Is a Learning Society?* edited by Paul Oliver, 31–45. Brookfield, VT: Ashgate, 1999.

Edwards, Richard, and Robin Usher. "Globalization, Diaspora Space and Pedagogy." Paper presented at the annual meeting of the American Educational Research Association, San Diego, April 1998.

Ehrenreich, Barbara. *Nickel and Dimed: On (Not) Getting by in America*. New York: Henry Holt, 2001.

Eppler, Marion A., and Beverly Harju, L. "Achievement Motivation Goals in Relation to Academic Performance in Traditional and Nontraditional College Students." *Research in Higher Education* 38, no. 5 (1997): 557–73.

ERIC Development Team. *Two-Year College Students: A Statistical Profile*. Los Angeles: ERIC Clearinghouse for Junior Colleges, 1982.

Fitzsimons, Patrick. "Changing Conceptions of Globalization: Changing Concepts of Education." *Educational Theory* 50, no. 4 (2000): 505–21.

Florida, Richard. *The Rise of the Creative Class: And How It's Transforming Work, Leisure, Community, and Everyday Life*. New York: Basic Books, 2002.

Floyd, Deborah L., Michael L. Skolnik, and Kenneth P. Walker, eds. *The Community College Baccalaureate: Emerging Trends and Policy Issues*. Sterling, VA: Stylus, 2004.

Freire, Paulo. *Pedagogy of the Oppressed*. New York: Continuum, 1994.

Frye, John. "Educational Paradigms in the Professional Literature of the Community College." In *Higher Education: Handbook of Theory And Research*, edited by John Smart, 181–224. New York: Agathon, 1994.

Frye, Northrop. *Anatomy of Criticism*. Toronto, ON: McClelland and Stewart, 1966.

Fuller, Jack. *Continuing Education and the Community College.* Chicago: Nelson–Hall, 1979.

Gallacher, Jim, Beth Crossman, Jon Field, and Barbara Merrill. "Learning Careers and The Social Space: Exploring the Fragile Identities of Adult Returners in the New Further Education." *International Journal of Lifelong Education* 21, no. 6 (2002): 493–509.

Gee, James Paul. "Identity as an Analytical Lens for Research in Education." *Review of Research in Education* 25 (2001): 99–125.

Gee, James Paul, Glynda Hull, and Colin Lankshear. *The New Work Order: Behind the Language of the New Capitalism.* Boulder, CO: Westview, 1996.

Giroux, Henry. *The Terror of Neoliberalism.* Boulder, CO: Paradigm, 2004.

Glick, Jennifer E., and Michael J. White. "Postsecondary School Participation of Immigrant and Native Youth: The Role of Familial Resources and Educational Expectations." *Social Science Research* 33 (2004): 272–99.

Gonzalez, Kenneth. "Campus Culture and the Experience of Chicano Students in Predominantly White Colleges and Universities." Paper presented at the annual meeting of the Association for the Study of Higher Education, San Antonio, TX, November 18–21, 1999.

———. "Inquiry as a Process of Learning about the Other and the Self." *Qualitative Studies in Education* 14, no. 4 (2001): 543–62.

Gonzalez, Kenneth, Carla Stoner, and Jennifer Jovel. "Examining Opportunities for Latinas in Higher Education: Toward a College Opportunity Framework." Paper presented at the annual meeting of the Association for the Study of Higher Education, Richmond, VA, 2001.

Gough, Noel. "Globalization and Curriculum: Theorizing a Transnational Imaginary." Paper presented at the annual meeting of the American Educational Research Association, San Diego, 1998.

Gould, Eric. *The University in a Corporate Culture.* New Haven, CT: Yale University Press, 2003.

Gouthro, Patricia. "Education for Sale: At What Cost? Lifelong Learning and the Marketplace." *International Journal of Lifelong Education* 21, no. 4 (2002): 334–46.

Government Accountability Office. "Public Community Colleges and Technical Schools (No. GAO–05–04)." Washington, DC: United States GAO, 2004.

Grosz, Karen Sue. *The Hierarchical Approach to Shared Governance.* Academic Senate for California Community Colleges, 1988. Available at http://www.academicsenate.cc.ca.us/Publications/Search/originalresults.asp. Accessed on June 23, 2005.

Grubb, W. Norton. *Honored but Invisible: An Inside Look at Teaching in Community Colleges.* New York: Routledge, 1999.

———. "Learning and Earning in the Middle: The Economic Benefits of Sub-Baccalaureate Education." Occasional paper, Community College Research Center, Teachers College, New York, September 1998.

———. "Second Chances in Changing Times: The Role of Community Colleges in Advancing Low-Skilled Workers." In *Low-Wage Workers in the New Economy,* edited by Richard Kazis and Marc S. Miller, 283–306. Washington, DC: Urban Institute Press, 2001.

———. *Using Community Colleges to Reconnect Disconnected Youth.* Menlo Park, CA: William and Flora Hewlett Foundation, 2003.

Grubb, W. Norton, Noreen Badway, and Denise Bell. "Community Colleges and the Equity Agenda: The Potential of Noncredit Education." *The ANNALS of the American Academy of Political and Social Sciences* (March 2003): 218–40.

Grubb, W. Norton, Noreen Badway, Denise Bell, Debra Bragg, and Maxine Russman. *Workforce, Economic and Community Development: The Changing Landscape of the Entrepreneurial Community College.* Berkeley: National Center for Research in Vocational Education, University of California, 1997.

Grubb, W. Norton, and Marvin Lazerson. *The Education Gospel.* Cambridge, MA: Harvard University Press, 2004.

Hagedorn, Linda. *Traveling Successfully on the Community College Pathway: The Research and Findings of the Transfer and Retention of Urban Community College Students Project.* Los Angeles: University of Southern California, Rossier School of Education, 2006.

Haleman, Diana. "Great Expectations: Single Mothers in Higher Education." *International Journal of Qualitative Studies in Education* 17, no. 6 (2004): 769–84.

Hasenfeld, Yeheskel. *Human Service Organizations.* Englewood Cliffs, NJ: Prentice Hall, 1983.

Held, David, and Anthony McGrew. "Globalization and the Liberal Democratic State." *Government and Opposition: The International Journal of Comparative Politics* 23, no. 2 (1993): 261–88.

Held, David, Anthony McGrew, David Goldblatt, and Jonathan Perraton. *Global Transformations: Politics, Economics and Culture.* Stanford, CA: Stanford University Press, 1999.

Herideen, Penelope E. *Policy, Pedagogy and Social Inequality: Community College Student Realities in Postindustrial America.* Westport, CT: Bergin and Garvey, 1998.

Higgins, Gina O'Connell. *Resilient Adults: Overcoming a Cruel Past.* San Francisco: Jossey-Bass, 1994.

Holland, Dorothy C., and Margaret A. Eisenhart. *Educated in Romance: Women, Achievement, and College Culture.* Chicago: University of Chicago Press, 1990.

Holland, Dorothy C., William Lachicotte, Debra Skinner, and Carole Cain. *Identity and Agency in Cultural Worlds.* Cambridge, MA: Harvard University Press, 1998.

Horn, Laura J., and C. Dennis Carroll. *Nontraditional Undergraduates: Trends in Enrollment from 1986–1992 and Persistence and Attainment among 1989–1990 Beginning Postsecondary Students.* U.S. Department of Education, Office of Educational Research and Improvement, NCES 97–578, National Center for Education Statistics, 1996.

Hudson, Lisa. "Demographic Attainment Trends in Postsecondary Education." In *The Knowledge Economy and Postsecondary Education,* edited by P. A. Graham and N. Stacey, 13–54. Washington, DC: National Academy Press, 2002.

Jacobs, Jerry A., and Sarah Winslow. "Welfare Reform and Enrollment in Postsecondary Education." *The ANNALS of the American Academy of Political and Social Sciences* (March 2003): 194–217.

Jenkins, Davis, and Katherine Boswell. *State Policies on Community College Workforce Development.* Denver: Education Commission of the States, Center for Community College Policy, 2002.

Johnson, Steven Lee. "Organizational Structures and the Performance of Contract Training Operations in American Community Colleges." PhD diss., University of Texas at Austin, 1995.

Kasworm, Carol E. "Adult Student Identity in an Intergenerational Community College Classroom." *Adult Education Quarterly* 56, no. 1 (2005): 3–20.

Kater, Sue, and John Levin. "Shared Governance in Community Colleges in the Global Economy." *Community College Journal of Research and Practice* 29, no. 1 (2005): 1–24.

Kempner, Ken. "The Community College as a Marginalized Institution." Unpublished paper presented at annual meeting of Association of the Study of Higher Education, Boston, 1991.

Kent, Susan E., and Michael J. Gimmestad. "Adult Undergraduate Student Persistence and Their Perceptions of How They Matter to the Institution." Paper presented at the Association for the Study of Higher Education, Kansas City, MO, 2004.

Keyman, E. Fuat. *Globalization, State, Identity, and Deference.* Atlantic Highlands, NJ: Humanities Press, 1997.

Kienzl, Gregory S. "The Triple Helix of Education and Earnings: The Effect of Schooling, Work, and Pathways onthe Economic Outcomes of Community College Students." Paper presented at the Association for the Study of Higher Education, Kansas City, MO, 2004.

Kingfisher, Catherine. *Western Welfare in Decline: Globalization and Women's Poverty.* Philadelphia: University of Pennsylvania Press, 2002.

Labaree, David F. "From Comprehensive High School to Community College: Politics, Markets, and the Evolution of Educational Opportunity." *Research in Sociology of Education and Socialization* 9 (1990): 203–240.

———. *How to Succeed in School Without Really Learning.* New Haven, CT: Yale University Press, 1997.

———. "Public Goods, Private Goods: The American Struggle over Educational Goals." *American Educational Research Journal* 34, no. 1 (1997): 39–81.

Laxer, Gordon. "Social Solidarity, Democracy, and Global Capitalism." *The Canadian Review of Sociology and Anthropology* (August 1995): 287–312.

Le Compte, Margaret, and Judith Preissle. *Ethnography and Qualitative Design in Educational Research.* San Diego, CA: Academic Press, 1993.

Levey, Tania. "Reexamining Community College Effects: New Techniques, New Outcomes." Paper presented at the Association for the Study of Higher Education, Philadelphia, November 2005.

Levin, John S. "The Business Culture of the Community College: Students as Consumers; Students as Commodities." In "Arenas of Entrepreneurship: Where Nonprofit and For-Profit Institutions Compete," edited by Brian Pusser. Special issue, *New Directions for Higher Education* 129 (2005): 11–26.

———. "The Community College as a Baccalaureate-Granting Institution." *The Review of Higher Education* 28, no. 1 (2004): 1–22.

———. *Globalizing the Community College: Strategies for Change in the Twenty-First Century.* New York: Palgrave, 2001.

———. *The Higher Credential.* Tucson, AZ: Canadian Embassy in Washington, DC, 2001.

———. "Missions and Structures: Bringing Clarity to Perceptions about Globalization and Higher Education in Canada." *Higher Education* 37, no. 4 (1999): 377–99.

———. "Neoliberal Policies and Community College Faculty Work." In *Handbook of Higher Education,* vol. 22, edited by John Smart and William Tierney. Norwell, MA: Kluwer Academic Publishers, 2007; 451–469.

———. "New World College: Community Colleges in the New Economy." Paper presented at the Canadian Society for the Study of Higher Education, Toronto, ON, May 2006.

———. "Nontraditional Students and Community Colleges: The Conflict of Justice and Neoliberalism. An Overview." Paper presented at the American Educational Research Association, Montreal, April 2005.

————. "Nouveau College: Community Colleges in the New Economy." Paper presented at the American Educational Research Association, San Francisco, April 2006.

————. "Public Policy, Community Colleges, and the Path to Globalization." *Higher Education* 42, no. 2 (2001): 237–62.

————. "The Revised Institution: The Community College Mission at the End of the Twentieth Century." *Community College Review* 28, no. 2 (2000): 1–25.

————. "Student Markets: The Business Culture of the Community College." Paper presented at the Association for the Study of Higher Education, Sacramento, CA, November 2002.

Levin, John, and John Dennison. "Responsiveness and Renewal in Canada's Community Colleges: A Study of Organizations." *The Canadian Journal of Higher Education* 19, no. 2 (1989): 41–57.

Levin, John S., Susan Kater, and Richard Wagoner. *Community College Faculty: At Work in the New Economy.* New York: Palgrave Macmillan, 2006.

Levin, John S., and Jeremy Levin. *The Costs of College.* DVD. Riverside, CA, 2006.

Lipset, Seymour. *American Exceptionalism: A Double-Edged Sword.* New York: W. W. Norton, 1996.

————. *Continental Divide: The Values and Institutions of the United States and Canada.* New York: Routledge, 1989.

Lipsky, Martin. *Street-Level Bureaucracy.* New York: Russell Sage Foundation, 1980.

London, Rebecca A. *The Role of Postsecondary Education in Welfare Recipients' Paths to Self-Sufficiency.* Santa Cruz: University of California, 2004.

March, James, and Michael Cohen. *Leadership and Ambiguity: The American College President.* New York: McGraw-Hill, 1974.

Marginson, Simon. *Educating Australia: Government, Economy, and Citizen since 1960.* Melbourne: Cambridge University Press, 1997.

Marginson, Simon, and Mark Considine. *The Enterprise University: Power, Governance, and Reinvention in Australia.* New York: Cambridge University Press, 2000.

Marshall, Catherine, and Gretchen Rossman. *Designing Qualitative Research*, 3rd ed. Thousand Oaks, CA: Sage, 1999.

Mason, Jennifer. *Qualitative Researching.* Thousand Oaks, CA: Sage, 1996.

Massey, Douglas S., Camille Z. Charles, Garvey F. Lundy, and Mary J. Fischer. *The Source of the River: The Social Origins of Freshmen at America's Selective Colleges and Universities.* Princeton, NJ: Princeton University Press, 2003.

Matus-Grossman, Lisa, and Susan Gooden. "Opening Doors: Students' Perspectives on Juggling Work, Family, and College." Manpower Demonstration Research (MDRC), 2002.

————. "Opening Doors to Earning Credentials: Impressions of Community College Access and Retention from Low-Wage Workers." Paper presented at the annual research conference of the Association for Public Policy Analysis and Management, Washington, DC, November 2001.

Maxwell, William, Linda Hagedorn, Hye Moon, Phillip Brocato, Kelly Wahl, and George Prather. "Community and Diversity in Urban Community Colleges: Coursetaking among Entering Students." *Community College Review* 30, no. 4 (2003): 21–46.

Mazzeo, Christopher, Sara Rab, and Susan Eachus. "Work-First or Work-Study: Welfare Reform, State Policy, and Access to Postsecondary Education." *The ANNALS of the American Academy of Political and Social Sciences* March (2003): 144–71.

McDonough, Patricia. *Choosing Colleges: How Social Class and Schools Structure Opportunity.* Albany: State University of New York Press, 1997.

McGrath, Dennis, and Martin Spear. *The Academic Crisis of the Community College.* Albany: State University of New York Press, 1991.

Meier, Ken. "The Community College Mission: History and Theory." Unpublished paper, Bakersfield, CA, 2004.

———. "The Community College Mission and Organizational Behavior." Unpublished paper. The Center for the Study of Higher Education, Tucson, Arizona, 1999.

———. "Social and Educational Origins of the Community College Movement: 1930–1945." Unpublished paper. Bakersfield, CA, 2004.

Meyer, John, and Brian Rowan. "Institutionalized Organizations: Formal Structure as Myth and Ceremony." *American Journal of Sociology* 83 (1977): 340–63.

Miles, Matthew, and A. Michael Huberman. *Qualitative Data Analysis.* Thousand Oaks, CA: Sage, 1994.

Mingle, James, Bruce Chaloux, and Angela Birkes. *Investing Wisely in Adult Learning Is Key to State Prosperity.* Atlanta: Southern Regional Education Board, 2005.

Mintzberg, Henry. *Mintzberg on Management: Inside Our Strange World of Organizations.* New York: Free Press, 1989.

———. *Power in and around Organizations.* Englewood Cliffs, NJ: Prentice Hall, 1983.

———. *Rise and Fall of Strategic Planning.* New York: Free Press, 1994.

Moffat, Michael. *Coming of Age in New Jersey: College and American Culture.* New Brunswick, NJ: Rutgers University Press, 1989.

Morest, Vanessa Smith. *The Role of Community Colleges in State Adult Education Systems: A National Analysis.* New York: Council for Advancement of Adult Literacy, 2004. Available at http://www.caalusa.org/columbiawp3.pdf.

Morgan, Gareth. *Images of Organization.* Thousand Oaks, CA: Sage, 1997.

Nagler, Alisa. "The Impact of Community College Education on the Nursing Shortage Crisis." Unpublished paper, Raleigh: North Carolina State University, 2005.

Nathan, Rebekah. *My Freshman Year: What a Professor Learned by Becoming a Student.* Ithaca, NY: Cornell University Press, 2005.

National Center for Education Statistics. "The Condition of Education 2000." Washington, DC: U.S. Department of Education, 2000.

———. "The Condition of Education 2001." Washington, DC: U.S. Department of Education, 2001.

———. "The Condition of Education 2002." Washington, DC: U.S. Department of Education, 2002.

———. "The Condition of Education 2003." Washington, DC: U.S. Department of Education, 2003.

———. "The Condition of Education 2004." Washington, DC: U.S. Department of Education, 2004.

———. "Percentage Distribution of Undergraduates, by Age and Their Average and Median Age (as of 12/31/99): 1999–2000." Washington, DC: U.S. Department of Education, 2000.

———. "Percentage of 1999–2000 Undergraduates with Various Risk Characteristics, and the Average Number of Risk Factors." Washington, DC: U.S. Department of Education, 2000.

Newman, Lynn. "Postsecondary Education Participation of Youth with Disabilities." In *After High School: A First Look at the Postschool Experiences of Youth with Disabilities. A Report from the National Longitudinal Transition Study-2 (NLTS2),* edited by Mary Wagner, Lynn Newman, Renee Cameto, Nicolle Garza and Phyllis Levine, 39–48. Menlo Park, CA: SRI, 2005.

O'Banion, Terry. *The Learning College for the Twenty-First Century.* Phoenix, AZ: American Council on Education and the Oryx Press, 1997.

O'Banion, Terry et al. *Teaching and Learning in the Community College.* Washington, DC: Community College Press, 1995.

Ogren, Christine A. "Rethinking the "Nontraditional" Student from a Historical Perspective: State Normal Schools in the Late-Nineteenth and Early-Twentieth Centuries." *The Journal of Higher Education* 74, no. 6 (2003): 640–64.

Olssen, Mark. *The Neoliberal Appropriation of Tertiary Education Policy: Accountability, Research, and Academic Freedom,* 2000. Available at http://www.surrey.ac.uk/Education/profiles/olssen/neo-2000.htm. Accessed in May 2004.

Pascarella, Ernest T., Christopher T. Pierson, Gregory C. Wolniak, and Patrick T. Terenzini. "First-Generation College Students: Additional Evidence on College Experiences and Outcomes." *The Journal of Higher Education* 75, no. 3 (2004): 249–84.

Pascarella, Ernest T., and Patrick T. Terenzini. *How College Affects Students.* San Francisco: Jossey Bass, 1991.

———. *How College Affects Students: A Third Decade of Research.* Vol. 2. San Francisco: Jossey Bass, 2005.

Perna, Laura W. "Differences in the Decision to Attend College among African Americans, Hispanics, and Whites." *The Journal of Higher Education* 71 (2000): 117–41.

Peter, Katharin, and Laura Horn. *Gender Differences in Participation and Completion of Undergraduate Education and How They Have Changed Over Time.* Washington, DC: U.S. Department of Education, National Center for Education Statistics, U.S. Government Printing Office, 2005.

Phillippe, Kent A., and Leila Gonzalez Sullivan. *National Profile of Community Colleges: Trends and Statistics, 4th Edition.* Washington, DC: American Association of Community Colleges, 2005

Phillippe, Kent A., and Madeline Patton. *National Profile of Community Colleges: Trends and Statistics,* 3rd ed. Washington, DC: Community College Press, American Association of Community Colleges, 2000.

Phillippe, Kent A., and Michael J. Valiga. *Faces of the Future: A Portrait of America's Community College Students.* Washington DC: American Association of Community Colleges, 2000.

Price, Derek V. "Defining the Gaps: Access and Success at America's Community Colleges." In *Keeping America's Promise,* 35–37. Indianapolis, IN: Lumina Foundation, 2004.

Prince, David, and Davis Jenkins. *Building Pathways to Success for Low-Skill Adult Students: Lessons for Community College Policy and Practice from a Statewide Longitudinal Tracking Study.* New York: Community College Research Center, Teachers College, Columbia University, 2005.

Puiggrós, Adriana. *Neoliberalism and Education in the Americas.* Boulder, CO: Westview, 1999.

Pusser, Brian. "Beyond Baldridge: Extending the Political Model of Higher Education Organization and Governance." *Educational Policy* 17, no. 1 (2003): 121–40.

———. *Burning Down the House: Politics, Governance, and Affirmative Action at the University Of California.* Albany: State University of New York Press, 2004.

Pusser, Brian, and Andrea Spreter. "A Framework for a Meta-analysis of Research on Adult Learners." Unpublished paper, Charlottesville, VA: University of Virginia, 2005.

Ratcliff, James. "Seven Streams in the Historical Development of the Modern Community College." In *A Handbook on the Community College in America*, edited by George Baker, 3–16. Westport, CT: Greenwood, 1994.

Rawls, John. *Political Liberalism*. New York: Columbia University Press, 1993.

———. *A Theory of Justice*. Cambridge, MA: Belknap Press of Harvard University Press, 1999.

Readings, Bill. *The University in Ruins*. Cambridge, MA: Harvard University Press, 1997.

Reich, Robert. "Hire Education." *Rolling Stone* (October 20, 1994): 119–25.

———. *The Work of Nations: Preparing Ourselves for Twenty-First Century Capitalism*. New York: Vintage Books, 1992.

Renner, K. Edward. "Racial Equity and Higher Education." *Academe* (January–February 2003): 38–43.

Rhoades, Gary, and Sheila Slaughter. "Academic Capitalism, Managed Professionals, and Supply-Side Higher Education." *Social Text* 15, no. 2 (1997): 9–38.

Rhoads, Robert, and James Valadez. *Democracy, Multiculturalism, and the Community College*. New York: Garland, 1996.

Richardson, Richard, and Louis Bender. *Fostering Minority Access and Achievement in Higher Education*. San Francisco: Jossey-Bass, 1987.

Richardson, Richard, Elizabeth Fisk, and Morris Okun. *Literacy in the Open-Access College*. San Francisco: Jossey-Bass, 1983.

Rifkin, Jeremy. *The End of Work: The Decline of the Global Labor Force and the Dawn of the Post-Market Era*. New York: G.P. Putnam's Sons, 1995.

Robertson, Roland. *Globalization: Social Theory and Global Culture*. London: Sage, 1992.

Roe, Mary Ann. *Education and U.S. Competitiveness: The Community College Role*. Austin: IC2 Institute, University of Texas, 1989.

Ross, Andrew. *No-Collar: The Humane Workplace and Its Hidden Costs*. New York: Basic Books, 2003.

Roueche, John, and George A. Baker III. *Access and Excellence*. Washington, DC: Community College Press, 1987.

Roueche, John E., Eileen E. Ely, and Suanne D. Roueche. *In Pursuit of Excellence: The Community College of Denver*. Washington, DC: Community College Press, 2001.

Roueche, John, and Suanne Roueche. *Between a Rock and a Hard Place: The At-Risk Student in the Open-Door College*. Washington, DC: Community College Press, 1993.

———. *High Stakes, High Performance: Making Remedial Education Work*. Washington, DC: Community College Press, 1999.

Roueche, John, Lynn Taber, and Suanne Roueche. *The Company We Keep: Collaboration in the Community College*. Washington, DC: American Association of Community Colleges, 1995.

Santos, Minerva. "The Motivations of First-Semester Hispanic Two-Year College Students." *Community College Review* 32, no. 3 (2004): 18–34.

Saul, John Ralston. *The Unconscious Civilization*. Concord, ON: House of Anansi Press, 1995.

Schwartz, Wendy. *Immigrants and Their Educational Attainment: Some Facts and Findings*. Washington, DC: ERIC Digest 116 (ED 402398), Office of Educational Research and Improvement, 1996.

Scott, J. *A Matter of Record*. Cambridge, UK: Polity Press, 1990.

Scott, W. Richard. *Institutions and Organizations*. Thousand Oaks, CA: Sage, 1995.

Seidman, Alan, ed. *College Student Retention: Formula for Student Success*. Westport, CT: Greenwood, 2005.

Sfard, Anna, and Anna Prusak. "Telling Identities: In Search of an Analytic Tool for Investigating Learning as a Culturally Shaped Activity." *Educational Researcher* 34, no. 4 (2005): 14–22.

Shaw, Kathleen M.. "Defining the Self: Construction of Identity in Community College Students." In *Community Colleges as Cultural Texts*, edited by Kathleen Shaw, James Valadez, and Robert Rhoads, 153–71. Buffalo: State University of New York Press, 1999.

Shaw, Kathleen M., Sara Goldrick-Rab, Christopher Mazzeo, and Jerry A. Jacobs. "Putting Poor People to Work: How the Work-First Ideology Eroded College Access for the Poor." Unpublished manuscript, 2005.

Shaw, Kathleen M., and Howard London. "Culture and Ideology in Keeping Transfer Commitment: Three Community Colleges." *The Review of Higher Education* 25, no. 1 (2001): 91–114.

Shaw, Kathleen M., and Sara Rab. "Market Rhetoric versus Reality in Policy and Practice: The Workforce Investment Act and Access to Community College Education and Training." *The ANNALS of the American Academy of Political and Social Science* (March 2003): 172–93.

Shaw, Kathleen M., Robert Rhoads, and James Valadez. "Community Colleges as Cultural Texts: A Conceptual Overview." In *Community Colleges as Cultural Texts*, edited by Kathleen Shaw, James Valadez, and Robert Rhoads, 1–13. Albany: State University of New York Press, 1999.

———, eds. *Community Colleges as Cultural Texts*. Albany: State University of New York Press, 1999.

Shupe, David. "Productivity, Quality, and Accountability in Higher Education." *Journal of Continuing Higher Education* (Winter 1999): 2–13.

Silva, Tim, Margaret Cahalan, Natalie Lacireno-Paquet, and Mathematica Policy Research Inc. "Adult Education Participation Decisions and Barriers: Review of Conceptual Frameworks and Empirical Studies." Prepared for U.S. Department of Education, Office of Educational Research and Development, National Center for Education Statistics, August 1998.

Skolnik, Michael. "The Virtual University and the Professoriate." In *The University in Transformation: Global Perspective on the Futures of the University*, edited by Sohail Inayatullah and Jennifer Gidley, 55–67. Westport, CT: Bergin and Garvey, 2000.

Slaughter, Sheila. *The Higher Learning*. Buffalo: State University of New York Press, 1990.

———. "Who Gets What and Why in Higher Education? Federal Policy and Supply-Side Institutional Resource Allocation." Presidential address at the annual meeting of the Association for the Study of Higher Education, Memphis, TN, 1997.

Slaughter, Sheila, and Larry Leslie. *Academic Capitalism, Politics, Policies, and the Entrepreneurial University*. Baltimore: Johns Hopkins University Press, 1997.

Slaughter, Sheila, and Gary Rhoades. *Academic Capitalism and the New Economy: Markets, State, and Higher Education*. Baltimore: Johns Hopkins University Press, 2004.

———. "The Neoliberal University." *New Labor Forum* (Spring/Summer 2000): 73–79.

Smith, Vicki. *Crossing the Great Divide: Worker Risk and Opportunity in the New Economy*. New York: Cornell University Press, 2001.

Sperber, Murray. *Beer and Circus: How Big-Time Sports Is Crippling Undergraduate Education*. New York: Henry Holt, 2000.

Stanton-Salazar, Ricardo D. "A Social Capital Framework for Understanding the Socialization of Racial Minority Children and Youth." *Harvard Educational Review* 67 (1997): 1–29.

Steele, Claude. "Expert Report of Claude M. Steele: Gratz et al. v. Bollinger et al., No. 97-75321 (E.D. Mich.); Grutter et al. v. Bollinger et al., No. 97-75928 (E.D. Mich.)."

Available at http://www.umich.edu/urel/admissions/legal/expert/steele.html. Accessed in 2003.

Stromquist, Nelly P. *Education in a Globalized World: The Connectivity of Economic Power, Technology, and Knowledge*. Lanham, MD: Rowman and Littlefield, 2002.

Surette, Brian. "Transfer from Two-Year to Four-Year College: An Analysis of Gender Differences." *Economics of Education Review* 20, no. 2 (2001): 151–63.

Swail, Watson Scott, Alberto F. Cabrera, Chul Lee, and Adriane Williams. *Latino Students and Educational Pipeline: A Three-Part Series*. Educational Policy Institute, 2005.

Szelényi, Katalin, and June C. Chang. "Educating Immigrants: The Community College Role." *Community College Review* 30, no. 2 (2002): 55–73.

Taber, Lynn. "Chapter and Verse: How We Came to Be Where We Are." In *The Company We Keep: Collaboration in the Community College*, edited by John Roueche, Lynn Taber, and Suanne Roueche, 25–37. Washington, DC: American Association of Community Colleges, 1995.

Tapscott, Don. *The Digital Economy: Promise and Peril in the Age of Networked Intelligence*. New York: McGraw Hill, 1996.

Taylor, Charles. *The Ethics of Authenticity*. Cambridge, MA: Harvard University Press, 1991.

Teeple, Gary. *Globalization and the Decline of Social Reform*. Atlantic Highlands, NJ: Humanities Press, 1995.

Terenzini, Patrick T., Leonard Springer, Patricia M. Yaeger, Ernest T. Pascarella, and Amaury Nora. "First-Generation College Students: Characteristics, Experiences, and Cognitive Development." *Research in Higher Education* 37, no. 1 (1996): 1–22.

Thomas, Veronica G. "Educational Experiences and Transitions of Reentry College Women: Special Consideration for African American Female Students." *Journal of Negro Education* 70, no. 3 (2001): 139–55.

Tinto, Vincent, Anne Goodsell Love, and Pat Russo. *Building Learning Communities for New College Students*. State College, PA: National Center on Postsecondary Teaching, Learning, and Assessment, 1994.

Torres, Carlos A., and Daniel Schugurensky. "The Political Economy of Higher Education in the Era of Neoliberal Globalization: Latin America in Comparative Perspective." *Higher Education* 43 (2002): 429–55.

Touraine, Alain. *Beyond Neoliberalism*. Translated by David Macey. Malden, MA: Blackwell, 2001.

Twombly, Susan, and Barbara Townsend. "Conclusion: The Future of Community Policy in the Twenty-First Century." In *Community Colleges: Policy in the Future Context*, edited by Barbara Townsend and Susan Twombly, 283–98. Westport, CT: Ablex, 2001.

Usher, Robin. "Identity, Risk, and Lifelong Learning." In *Lifelong and Continuing Education: What Is a Learning Society*, edited by Paul Oliver, 65–82. Brookfield, VT: Ashgate, 1999.

Valadez, James. "Cultural Capital and Its Impact on the Aspirations of Nontraditional Community College Students." *Community College Review* 21, no. 3 (1996): 30–44.

———. "Searching for a Path out of Poverty: Exploring the Achievement Ideology of a Rural Community College." *Adult Education Quarterly* 50, no. 3 (2000): 212–30.

Vaughan, George. "The Big Squeeze at Community Colleges." *The News & Observer*, March 24, 2002.

———. *The Community College Story*. Washington, DC: American Association of Community Colleges, 2000.

Voorhees, Richard, and P. Lingenfelter. *Adult Learning and State Policy*. Chicago: State Higher Education Executive Officers and Council for Adult and Experiential Learning, 2003.

Wagner, Mary. "Characteristics of Out-of-School Youth with Disabilities." In *After High School: A First Look at the Postschool Experiences of Youth with Disabilities. A Report from the National Longitudinal Transition Study-2 (NLTS2)*, edited by Mary Wagner, Lynn Newman, Renee Cameto, Nicolle Garza, and Phyllis Levine, chapter 2. Menlo Park, CA: SRI, 2005. Available at www.nlts2.org/pdfs/afterhighschool_chp2.pdf.

Wagoner, Richard. L, Amy Scott Metcalfe, and Israel Olaore. "Fiscal Reality and Academic Quality: Part-Time Faculty and the Challenge to Organizational Culture at Community Colleges." *Community College Journal of Research and Practice* 29 (2005): 1–20.

Wain, Kenneth. "The Learning Society: Postmodern Politics." *International Journal of Lifelong Education* 19, no. 1 (2000): 36–53.

Waldinger, Roger, and Michael Lichter, I. *How the Other Half Works: Immigration and the Social Organization of Labor*. Berkeley: University of California Press, 2003.

Walters, Shirley, and Kathy Watters. "Lifelong Learning, Higher Education, and Active Citizenship: From Rhetoric to Action." *International Journal of Lifelong Education* 20, no. 6 (2001): 471–78.

Waters, Malcolm. *Globalization*. New York: Routledge, 1996.

Weick, Karl. "Educational Organizations as Loosely Coupled Systems." *Administrative Science Quarterly* 21 (1976): 1–19.

Weis, Lois. *Between Two Worlds: Black Students in an Urban Community College*. Boston: Routledge and Kegan Paul, 1985.

Weissman, Julie, Carole Bulakowski, and Marci Jumisko. "A Study of White, Black, and Hispanic Students' Transition to a Community College." *Community College Review* 26, no. 2 (1998): 19–42.

Welch, Anthony P. "Globalisation, Postmodernity, and the State: Comparative Education Facing the Third Millennium." *Comparative Education* 37, no. 4 (2000): 475–92.

Wellman, Jane V. *State Policy and Community College-Baccalaureate Transfer*. National Center for Public Policy and Higher Education and the Institute for Higher Education Policy, 2002.

White, Kenneth. "Shared Governance in California." *New Directions for Community Colleges* 102 (1998): 19–29.

Wolfe, Tom. *I Am Charlotte Simmons*. New York: Farrar, Straus and Giroux, 2004.

Wolgemuth, Jennifer, Nathalie Kees, and Lynn Safarik. "A Critique of Research on Women Published in the Community College Journal of Research and Practice: 1990–2000." *Community College Journal of Research and Practice* 27, nos. 9–10 (2003): 757–68.

Appendix A

Institutional Student Characteristics

Wake Technical Community College

	Data	Source (see Source List)

Student Demographics

Total Enrollment (Fall 2002)	10,564	2
Male	46.2%	2
Female	53.8%	2
White, non-Hispanic	62.2%	2
Black, non-Hispanic	21.9%	2
Hispanic	2.3%	2
Asian or Pacific Islander	2.7%	2
American Indian or Alaskan Native	0.5%	2
Race/ethnicity unknown	1.9%	2
Nonresident alien	8.5%	2
Student Counts, 2001–2002:		
Full-time, first-time undergraduate	442	2
Percent full-time, first-time undergraduate	10.0%	2
Percent full-time, first-time under-graduate receiving financial aid	36.0%	2
Awards / Degrees Conferred (7/1/02–6/30/03)		
Associate degrees, total	862	3
Graduation Rates (cohort year 1999)		
Graduation rate within 150% of normal time to completion	6.9%	4
Transfer-out rate	n/a	4
Graduation rates by gender		
Men	5.4%	4
Women	8.5%	4
Graduation rates by race/ethnicity		
White, non-Hispanic	7.8%	4
Black, non-Hispanic	6.1%	4
Hispanic	0.0%	4
Asian/Pacific Islander	6.5%	4
American Indian/Alaska Native	n/a	4
Race/ethnicity unknown	0.0%	4
Nonresident alien	5.0%	4

	Data	**Source (see Source List)**

Finance

Full-time tuition and fees costs (2003–2004)

	Data	Source
In-district	n/a	1
In-state	$1,152.00	1
Out-of-state	$6,320.00	1

REVENUES BY SOURCE, FY 2001

Tuition and fees	$7,092,653	53
Federal appropriations	$712,960	53
State appropriations	$25,983,216	53
Local appropriations	$7,810,000	53
Federal grants and contracts	$2,925,242	53
State grants and contracts	$219,460	53
Local grants and contracts	$9,570	53
Private gifts, grants, and contracts	$80,834	53
Endowment income	$0	53
Sales and services of educational activities	$0	53
Auxiliary enterprises	$423,088	53
Hospital revenues	$0	53
Other sources	$283,079	53
Independent operations	$0	53
Total current funds revenues	$45,540,102	53

EXPENDITURES BY FUNCTION, FY 2001

Instruction	$22,733,011	53
Research	$0	53
Public service	$8,871	53
Academic support	$5,287,150	53
Student services	$3,246,435	53
Institutional support	$4,604,991	53
Operation and maintenance of plant	$4,147,745	53
Scholarships and fellowships	$2,934,385	53
Mandatory transfers	$0	53
Nonmandatory transfers	$0	53
Total educational and general expenditures	$42,962,588	53
Auxiliary enterprises	$206,154	53
Auxiliary enterprises (nonmandatory)	$0	53
Hospital expenditures	$0	53
Hospitals (nonmandatory)	$0	53
Independent operations	$0	53
Independent operations (nonmandatory)	$0	53
Other current funds expenditure	$0	53
Total current funds expenditures and transactions	$43,168,742	53
Amount salaries and wages, total E and G expenditures	$26,866,617	53
Employee fringe benefits, institutional	$4,622,266	53
E and G employee fringe benefits, paid from noninstitutional accounts	$0	53
Total E and G employee compensation	$31,488,883	53

	Data	**Source (see Source List)**

Other Institutional Data

Geographic Location	Wake County, NC	
Type of Institution	Public, 2-year	1
Degrees Awarded	Associate's	1
Carnegie Classification	Associate's College	1

Community Demographics

County	Wake	54
Population, 2000	627,846	54
Male	49.60%	54
Female	50.40%	54
Education Level		
County Percent High School Graduates, 2000	89.30%	54
County Percent Bachelor's or Higher, 2000	43.90%	54
Race and Ethnicity, 2000		
White	72.40%	54
Black or African American	19.70%	54
American Indian and Alaska Native	0.30%	54
Asian	3.40%	54
Native Hawaiian and Other Pacific Islander	<0.10%	54
Persons reporting some other race	2.50%	54
Persons reporting two or more races	1.60%	54
Persons of Hispanic or Latino origin	5.40%	54
White persons, not of Hispanic/ Latino origin	69.90%	54
Households, 2000	242,040	54
Persons per household, 2000	2.51	54
Median household income, 1999	$54,988	54
Per capita money income, 1999	$27,004	54
Persons below poverty, 1999	7.80%	54
Land Area, 2000 (square miles)	832.0	54
Persons per square mile, 2000	754.7	54
Nearest metropolitan area	Raleigh-Durham-Chapel Hill	54
Employment		
Private nonfarm establishments with paid employees, 2001	20,812	54
Private nonfarm employment, 2001	355,071	54
Minority-owned firms, percent of total, 1997	12.30%	54
Women-owned firms, percent of total, 1997	26.00%	54

Johnston Community College

	Data	Source (see Source List)

Student Demographics

Total Enrollment (Fall 2002)	3,651	6
Male	38.9%	6
Female	61.1%	6
White, non-Hispanic	73.9%	6
Black, non-Hispanic	21.5%	6
Hispanic	1.6%	6
Asian or Pacific Islander	0.6%	6
American Indian or Alaskan Native	0.7%	6
Race/ethnicity unknown	1.5%	6
Nonresident alien	0.3%	6
Student Counts, 2001–2002:		
Full-time, first-time undergraduate	311	6
Percent full-time, first-time undergraduate	9.0%	6
Percent full-time, first-time undergraduate receiving financial aid	37.0%	6
Awards / Degrees Conferred (7/1/02–6/30/03)		
Associate degrees, total	258	7
Graduation Rates (cohort year 1999)		
Graduation rate within 150% of normal time to completion	28.3%	8
Transfer-out rate	5.8%	8
Graduation rates by gender		
Men	34.9%	8
Women	21.1%	8
Graduation rates by race/ethnicity		
White, non-Hispanic	21.5%	8
Black, non-Hispanic	43.9%	8
Hispanic	n/a	8
Asian/Pacific Islander	n/a	8
American Indian/Alaska Native	n/a	8
Race/ethnicity unknown	n/a	8
Nonresident alien	n/a	8

Finance

Full-time tuition and fees costs (2003–2004)		
In-district	n/a	5
In-state	$1,206.00	5
Out-of-state	$6,374.00	5
REVENUES BY SOURCE, FY 2001		
Tuition and fees	$1,703,914	53
Federal appropriations	$0	53
State appropriations	$11,347,949	53
Local appropriations	$1,800,000	53
Federal grants and contracts	$1,688,088	53

	Data	Source (see Source List)
State grants and contracts	$858,430	53
Local grants and contracts	$255,524	53
Private gifts, grants, and contracts	$49,984	53
Endowment income	$0	53
Sales and services of educational activities	$0	53
Auxiliary enterprises	$1,171,049	53
Hospital revenues	$0	53
Other sources	$195,896	53
Independent operations	$0	53
Total current funds revenues	$19,070,834	53

EXPENDITURES BY FUNCTION, FY 2001

Instruction	$8,539,599	53
Research	$0	53
Public service	$0	53
Academic support	$1,703,156	53
Student services	$928,799	53
Institutional support	$3,223,953	53
Operation and maintenance of plant	$1,526,266	53
Scholarships and fellowships	$2,085,740	53
Mandatory transfers	$0	53
Nonmandatory transfers	-$2,625	53
Total educational and general expenditures	$18,004,888	53
Auxiliary enterprises	$1,171,702	53
Auxiliary enterprises (nonmandatory)	$0	53
Hospital expenditures	$0	53
Hospitals (nonmandatory)	$0	53
Independent operations	$0	53
Independent operations (nonmandatory)	$0	53
Other current funds expenditure	$2,624	53
Total current funds expenditures and transactions	$19,179,214	53
Amount salaries and wages, total E and G expenditures	$10,665,620	53
Employee fringe benefits, institutional	$2,093,503	53
E and G employee fringe benefits, paid from noninstitutional accounts	$0	53
Total E and G employee compensation	$12,759,123	53

Other Institutional Data

Geographic Location	Johnston County, NC	
Type of Institution	Public, 2-year	5
Degrees Awarded	Associate's	5
Carnegie Classification	Associate's College	5

Community Demographics

County	Johnston	55
Population, 2000	121,965	55
Male	49.70%	55
Female	50.30%	55

Education Level

County Percent High School Graduates, 2000	75.90%	55
County Percent Bachelor's or Higher, 2000	15.90%	55

	Data	Source (see Source List)
Race and Ethnicity, 2000		
White	78.10%	55
Black or African American	15.70%	55
American Indian and Alaska Native	0.40%	55
Asian	0.30%	55
Native Hawaiian and Other Pacific Islander	<0.10%	55
Persons reporting some other race	4.50%	55
Persons reporting two or more races	1.00%	55
Persons of Hispanic or Latino origin	7.70%	55
White persons, not of Hispanic/Latino origin	75.30%	55
Households, 2000	46,595	55
Persons per household, 2000	2.58	55
Median household income, 1999	$40,872	55
Per capita money income, 1999	$18,788	55
Persons below poverty, 1999	12.80%	55
Land Area, 2000 (square miles)	792.0	55
Persons per square mile, 2000	154.0	55
Nearest metropolitan area	Raleigh-Durham-Chapel Hill	55
Employment		
Private nonfarm establishments with paid employees, 2001	2,561	55
Private nonfarm employment, 2001	29,519	55
Minority-owned firms, percent of total, 1997	7.00%	55
Women-owned firms, percent of total, 1997	21.90%	55

Mountain View College

	Data	Source (see Source List)

Student Demographics

Total Enrollment (Fall 2002)	6,561	10
Male	39.6%	10
Female	60.4%	10
White, non-Hispanic	25.4%	10
Black, non-Hispanic	30.1%	10
Hispanic	37.5%	10
Asian or Pacific Islander	3.1%	10
American Indian or Alaskan Native	0.4%	10
Race/ethnicity unknown	1.4%	10
Nonresident alien	2.0%	10
Student Counts, 2001–2002:		
Full-time, first-time undergraduate	362	10
Percent full-time, first-time undergraduate	6.0%	10
Percent full-time, first-time undergraduate receiving financial aid	56.0%	10
Awards / Degrees Conferred (7/1/02–6/30/03)		
Associate degrees, total	314	11
Graduation Rates (cohort year 1999)		
Graduation rate within 150% of normal time to completion	9.2%	12
Transfer-out rate	30.3%	12
Graduation rates by gender		
Men	8.1%	12
Women	10.1%	12
Graduation rates by race/ethnicity		
White, non-Hispanic	1.7%	12
Black, non-Hispanic	8.6%	12
Hispanic	13.9%	12
Asian/Pacific Islander	insufficient data	12
American Indian/Alaska Native	n/a	12
Race/ethnicity unknown	n/a	12
Nonresident alien	n/a	12

Finance

Full-time tuition and fees costs (2003–2004)		
In-district	$840.00	9
In-state	$1,400.00	9
Out-of-state	$2,240.00	9
REVENUES BY SOURCE, FY 2001		
Tuition and fees	$3,976,936	53
Federal appropriations	$0	53
State appropriations	$8,595,741	53
Local appropriations	$6,706,601	53
Federal grants and contracts	$3,429,736	53

	Data	Source (see Source List)
State grants and contracts	$222,119	53
Local grants and contracts	$443,565	53
Private gifts, grants, and contracts	$0	53
Endowment income	$0	53
Sales and services of educational activities	$0	53
Auxiliary enterprises	$291,138	53
Hospital revenues	$0	53
Other sources	$50,859	53
Independent operations	$0	53
Total current funds revenues	$23,716,695	53

EXPENDITURES BY FUNCTION, FY 2001

	Data	Source
Instruction	$11,189,564	53
Research	$0	53
Public service	$290,454	53
Academic support	$1,507,778	53
Student services	$2,434,192	53
Institutional support	$2,422,370	53
Operation and maintenance of plant	$1,760,842	53
Scholarships and fellowships	$3,627,437	53
Mandatory transfers	$190,646	53
Nonmandatory transfers	$0	53
Total educational and general expenditures	$23,423,283	53
Auxiliary enterprises	$293,412	53
Auxiliary enterprises (nonmandatory)	-$311,021	53
Hospital expenditures	$0	53
Hospitals (nonmandatory)	$0	53
Independent operations	$0	53
Independent operations (nonmandatory)	$0	53
Other current funds expenditure	$0	53
Total current funds expenditures and transactions	$23,716,695	53
Amount salaries and wages, total E and G expenditures	$13,303,389	53
Employee fringe benefits, institutional	$346,835	53
E and G employee fringe benefits, paid from noninstitutional accounts	$1,035,300	53
Total E and G employee compensation	$14,685,524	53

Other Institutional Data

	Data	Source
Geographic Location	Dallas, TX	
Type of Institution	Public, 2-year	9
Degrees Awarded	Associate's	9
Carnegie Classification	Associate's College	9

Community Demographics

	Data	Source
County	Dallas	56
Population, 2000	2,284,096	56
Male	49.90%	56
Female	50.10%	56

	Data	Source (see Source List)
Education Level		
County Percent High School Graduates, 2000	75.00%	56
County Percent Bachelor's or Higher, 2000	27.00%	56
Race and Ethnicity, 2000		
White	58.40%	56
Black or African American	20.30%	56
American Indian and Alaska Native	0.60%	56
Asian	4.00%	56
Native Hawaiian and Other Pacific Islander	0.10%	56
Persons reporting some other race	14.00%	56
Persons reporting two or more races	2.70%	56
Persons of Hispanic or Latino origin	29.90%	56
White persons, not of Hispanic/Latino origin	44.30%	56
Households, 2000	807,621	56
Persons per household, 2000	2.71	56
Median household income, 1999	$43,324	56
Per capita money income, 1999	$22,603	56
Persons below poverty, 1999	13.40%	56
Land Area, 2000 (square miles)	880.0	56
Persons per square mile, 2000	2,522.6	56
Nearest metropolitan area	Dallas, TX	56
Employment		
Private nonfarm establishments with		
paid employees, 2001	63,613	56
Private nonfarm employment, 2001	1,505,640	56
Minority-owned firms, percent of total, 1997	22.30%	56
Women-owned firms, percent of total, 1997	25.80%	56

Pima Community College

	Data	Source (see Source List)

Student Demographics

Total Enrollment (Fall 2002)	31,135	14
Male	43.3%	14
Female	56.7%	14
White, non-Hispanic	52.8%	14
Black, non-Hispanic	3.8%	14
Hispanic	28.8%	14
Asian or Pacific Islander	3.1%	14
American Indian or Alaskan Native	3.3%	14
Race/ethnicity unknown	6.0%	14
Nonresident alien	2.1%	14
Student Counts, 2001–2002:		
Full-time, first-time undergraduate	1493	14
Percent full-time, first-time undergraduate	5.0%	14
Percent full-time, first-time undergraduate		
receiving financial aid	61.0%	14
Awards / Degrees Conferred (7/1/02–6/30/03)		
Associate degrees, total	1515	15
Graduation Rates (cohort year 1999)		
Graduation rate within 150% of normal		
time to completion	29.4%	16
Transfer-out rate	6.5%	16
Graduation rates by gender		
Men	19.2%	16
Women	16.4%	16
Graduation rates by race/ethnicity		
White, non-Hispanic	27.0%	16
Black, non-Hispanic	36.1%	16
Hispanic	32.0%	16
Asian/Pacific Islander	30.2%	16
American Indian/Alaska Native	29.2%	16
Race/ethnicity unknown	26.3%	16
Nonresident alien	insufficient data	16

Finance

Full-time tuition and fees costs (2003–2004)		
In-district	n/a	13
In-state	$1,018.00	13
Out-of-state	$5,074.00	13
REVENUES BY SOURCE, FY 2001		
Tuition and fees	$22,149,336	53
Federal appropriations	$0	53
State appropriations	$19,963,100	53
Local appropriations	$47,812,879	53
Federal grants and contracts	$22,103,154	53

	Data	Source (see Source List)
State grants and contracts	$196,144	53
Local grants and contracts	$0	53
Private gifts, grants, and contracts	$1,156,406	53
Endowment income	$0	53
Sales and services of educational activities	$0	53
Auxiliary enterprises	$1,239,756	53
Hospital revenues	$0	53
Other sources	$5,129,051	53
Independent operations	$0	53
Total current funds revenues	$119,749,826	53

EXPENDITURES BY FUNCTION, FY 2001

Instruction	$47,426,661	53
Research	$0	53
Public service	$0	53
Academic support	$15,862,442	53
Student services	$12,645,361	53
Institutional support	$17,147,466	53
Operation and maintenance of plant	$7,799,621	53
Scholarships and fellowships	$13,631,013	53
Mandatory transfers	$2,165,000	53
Nonmandatory transfers	-$694,578	53
Total educational and general expenditures	$115,982,986	53
Auxiliary enterprises	$1,147,974	53
Auxiliary enterprises (nonmandatory)	$0	53
Hospital expenditures	$0	53
Hospitals (nonmandatory)	$0	53
Independent operations	$0	53
Independent operations (nonmandatory)	$0	53
Other current funds expenditure	$0	53
Total current funds expenditures and transactions	$117,130,960	53
Amount salaries and wages, total E and G expenditures	$66,658,429	53
Employee fringe benefits, institutional	$10,664,347	53
E and G employee fringe benefits, paid from noninstitutional accounts	$0	53
Total E and G employee compensation	$77,322,776	53

Other Institutional Data

Geographic Location	Tucson, AZ	
Type of Institution	Public, 2-year	13
Degrees Awarded	Associate's	13
Carnegie Classification	Associate's College	13

Community Demographics

County	Pima	57
Population, 2000	892,798	57
Male	48.90%	57
Female	51.10%	57

Education Level

County Percent High School Graduates, 2000	83.40%	57
County Percent Bachelor's or Higher, 2000	26.70%	57

	Data	Source (see Source List)
Race and Ethnicity, 2000		
White	75.10%	57
Black or African American	3.00%	57
American Indian and Alaska Native	3.20%	57
Asian	2.00%	57
Native Hawaiian and Other Pacific Islander	0.10%	57
Persons reporting some other race	13.30%	57
Persons reporting two or more races	3.20%	57
Persons of Hispanic or Latino origin	29.30%	57
White persons, not of Hispanic/Latino origin	61.50%	57
Households, 2000	332,350	57
Persons per household, 2000	2.47	57
Median household income, 1999	$36,758	57
Per capita money income, 1999	$19,785	57
Persons below poverty, 1999	14.70%	57
Land Area, 2000 (square miles)	9,186.0	57
Persons per square mile, 2000	91.8	57
Nearest metropolitan area	Tucson, AZ	57
Employment		
Private nonfarm establishments		
with paid employees, 2001	18,828	57
Private nonfarm employment, 2001	293,987	57
Minority-owned firms, percent of total, 1997	16.70%	57
Women-owned firms, percent of total, 1997	27.90%	57

GateWay Community College

	Data	Source (see Source List)

Student Demographics

Total Enrollment (Fall 2002)	7,969	18
Male	50.3%	18
Female	49.7%	18
White, non-Hispanic	54.5%	18
Black, non-Hispanic	7.6%	18
Hispanic	22.5%	18
Asian or Pacific Islander	2.7%	18
American Indian or Alaskan Native	5.4%	18
Race/ethnicity unknown	6.7%	18
Nonresident alien	0.7%	18
Student Counts, 2001–2002:		
Full-time, first-time undergraduate	123	18
Percent full-time, first-time undergraduate	2.0%	18
Percent full-time, first-time undergraduate		
receiving financial aid	56.0%	18
Awards / Degrees Conferred (7/1/02–6/30/03)		
Associate degrees, total	259	19
Graduation Rates (cohort year 1999)		
Graduation rate within 150% of normal		
time to completion	16.4%	20
Transfer-out rate	12.3%	20
Graduation rates by gender		
Men	insufficient data	20
Women	13.0%	20
Graduation rates by race/ethnicity		
White, non-Hispanic	20.0%	20
Black, non-Hispanic	insufficient data	20
Hispanic	insufficient data	20
Asian/Pacific Islander	n/a	20
American Indian/Alaska Native	n/a	20
Race/ethnicity unknown	n/a	20
Nonresident alien	n/a	20

Finance

Full-time tuition and fees costs (2003–2004)		
In-district	$1,234.00	17
In-state	$5,050.00	17
Out-of-state	$5,194.00	17
REVENUES BY SOURCE, FY 2001		
Tuition and fees	$8,595,948	53
Federal appropriations	$0	53
State appropriations	$2,491,279	53
Local appropriations	$12,227,554	53
Federal grants and contracts	$2,504,315	53

	Data	Source (see Source List)
State grants and contracts	$1,585,436	53
Local grants and contracts	$37,136	53
Private gifts, grants, and contracts	$432,954	53
Endowment income	$0	53
Sales and services of educational activities	$0	53
Auxiliary enterprises	$351,474	53
Hospital revenues	$0	53
Other sources	$4,143,191	53
Independent operations	$0	53
Total current funds revenues	$32,369,287	53

EXPENDITURES BY FUNCTION, FY 2001

Instruction	$13,932,785	53
Research	$0	53
Public service	$16,416	53
Academic support	$1,909,373	53
Student services	$2,537,716	53
Institutional support	$4,149,790	53
Operation and maintenance of plant	$2,350,186	53
Scholarships and fellowships	$2,616,574	53
Mandatory transfers	$34,172	53
Nonmandatory transfers	$1,969,819	53
Total educational and general expenditures	$29,516,831	53
Auxiliary enterprises	$1,268,524	53
Auxiliary enterprises (nonmandatory)	$26,159	53
Hospital expenditures	$0	53
Hospitals (nonmandatory)	$0	53
Independent operations	$0	53
Independent operations (nonmandatory)	$0	53
Other current funds expenditure	$0	53
Total current funds expenditures and transactions	$30,785,355	53
Amount salaries and wages, total E and G expenditures	$16,679,266	53
Employee fringe benefits, institutional	$2,941,166	53
E and G employee fringe benefits, paid from noninstitutional accounts	$0	53
Total E and G employee compensation	$19,620,432	53

Other Institutional Data

Geographic Location	Phoenix, AZ	
Type of Institution	Public, 2-year	17
Degrees Awarded	Associate's	17
Carnegie Classification	Associate's College	17

Community Demographics

County	Maricopa	58
Population, 2000	3,389,260	58
Male	50.00%	58
Female	50.00%	58

Education Level

County Percent High School Graduates, 2000	82.50%	58
County Percent Bachelor's or Higher, 2000	25.90%	58

	Data	Source (see Source List)
Race and Ethnicity, 2000		
White	77.40%	58
Black or African American	3.70%	58
American Indian and Alaska Native	1.80%	58
Asian	2.20%	58
Native Hawaiian and Other Pacific Islander	0.10%	58
Persons reporting some other race	11.90%	58
Persons reporting two or more races	2.90%	58
Persons of Hispanic or Latino origin	24.80%	58
White persons, not of Hispanic/Latino origin	66.20%	58
Households, 2000	1,132,886	58
Persons per household, 2000	2.67	58
Median household income, 1999	$45,358	58
Per capita money income, 1999	$22,251	58
Persons below poverty, 1999	11.70%	58
Land Area, 2000 (square miles)	9,203.0	58
Persons per square mile, 2000	333.8	58
Nearest metropolitan area	Phoenix-Mesa, AZ	58
Employment		
Private nonfarm establishments		
with paid employees, 2001	73,102	58
Private nonfarm employment, 2001	1,377,298	58
Minority-owned firms, percent of total, 1997	11.10%	58
Women-owned firms, percent of total, 1997	26.30%	58

Edmonds Community College

	Data	Source (see Source List)

Student Demographics

Total Enrollment (Fall 2002)	8,224	22
Male	43.7%	22
Female	56.3%	22
White, non-Hispanic	68.7%	22
Black, non-Hispanic	3.9%	22
Hispanic	4.0%	22
Asian or Pacific Islander	10.4%	22
American Indian or Alaskan Native	1.3%	22
Race/ethnicity unknown	6.8%	22
Nonresident alien	4.9%	22
Student Counts, 2001–2002:		
Full-time, first-time undergraduate	556	22
Percent full-time, first-time undergraduate	7.0%	22
Percent full-time, first-time undergraduate receiving financial aid	32.0%	22
Awards / Degrees Conferred (7/1/02–6/30/03)		
Associate degrees, total	803	23
Graduation Rates (cohort year 1999)		
Graduation rate within 150% of normal time to completion	27.7%	24
Transfer-out rate	21.8%	24
Graduation rates by gender		
Men	23.2%	24
Women	32.9%	24
Graduation rates by race/ethnicity		
White, non-Hispanic	25.8%	24
Black, non-Hispanic	insufficient data	24
Hispanic	insufficient data	24
Asian/Pacific Islander	39.7%	24
American Indian/Alaska Native	n/a	24
Race/ethnicity unknown	n/a	24
Nonresident alien	insufficient data	24

Finance

Full-time tuition and fees costs (2003–2004)		
In-district	n/a	21
In-state	$2,187.00	21
Out-of-state	$7,359.00	21
REVENUES BY SOURCE, FY 2001		
Tuition and fees	$11,930,086	53
Federal appropriations	$0	53
State appropriations	$17,185,090	53
Local appropriations	$0	53
Federal grants and contracts	$6,733,327	53

	Data	Source (see Source List)
State grants and contracts	$6,232,024	53
Local grants and contracts	$1,232,878	53
Private gifts, grants, and contracts	$2,495,352	53
Endowment income	$0	53
Sales and services of educational activities	$1,147,185	53
Auxiliary enterprises	$6,607,383	53
Hospital revenues	$0	53
Other sources	$1,101,875	53
Independent operations	$0	53
Total current funds revenues	$54,665,200	53

EXPENDITURES BY FUNCTION, FY 2001

	Data	Source
Instruction	$25,758,645	53
Research	$0	53
Public service	$0	53
Academic support	$2,467,931	53
Student services	$2,997,183	53
Institutional support	$6,648,290	53
Operation and maintenance of plant	$3,735,009	53
Scholarships and fellowships	$5,762,539	53
Mandatory transfers	$0	53
Nonmandatory transfers	-$200,000	53
Total educational and general expenditures	$47,169,597	53
Auxiliary enterprises	$7,172,507	53
Auxiliary enterprises (nonmandatory)	$0	53
Hospital expenditures	$0	53
Hospitals (nonmandatory)	$0	53
Independent operations	$0	53
Independent operations (nonmandatory)	$0	53
Other current funds expenditure	$0	53
Total current funds expenditures and transactions	$54,342,104	53
Amount salaries and wages, total E and G expenditures	$24,018,602	53
Employee fringe benefits, institutional	$6,308,005	53
E and G employee fringe benefits, paid from noninstitutional accounts	$0	53
Total E and G employee compensation	$30,326,607	53

Other Institutional Data

Geographic Location	Lynnwood, WA	
Type of Institution	Public, 2-year	21
Degrees Awarded	Associate's	21
Carnegie Classification	Associate's College	21

Community Demographics

County	Snohomish	59
Population, 2000	639,409	59
Male	50.00%	59
Female	50.00%	59

Education Level

County Percent High School Graduates, 2000	89.20%	59
County Percent Bachelor's or Higher, 2000	24.40%	59

	Data	Source (see Source List)
Race and Ethnicity, 2000		
White	85.60%	59
Black or African American	1.70%	59
American Indian and Alaska Native	1.40%	59
Asian	5.80%	59
Native Hawaiian and Other Pacific Islander	0.30%	59
Persons reporting some other race	1.90%	59
Persons reporting two or more races	3.40%	59
Persons of Hispanic or Latino origin	4.70%	59
White persons, not of Hispanic/Latino origin	83.40%	59
Households, 2000	224,852	59
Persons per household, 2000	2.65	59
Median household income, 1999	$53,060	59
Per capita money income, 1999	$23,417	59
Persons below poverty, 1999	6.90%	59
Land Area, 2000 (square miles)	2,089.0	59
Persons per square mile, 2000	290.1	59
Nearest metropolitan area	Seattle-Bellevue-Everett, WA	59
Employment		
Private nonfarm establishments		
with paid employees, 2001	15,355	59
Private nonfarm employment, 2001	209,047	59
Minority-owned firms, percent of total, 1997	7.90%	59
Women-owned firms, percent of total, 1997	27.90%	59

Virginia Highlands Community College

	Data	Source (see Source List)

Student Demographics

Total Enrollment (Fall 2002)	2,595	26
Male	41.2%	26
Female	58.8%	26
White, non-Hispanic	97.8%	26
Black, non-Hispanic	1.4%	26
Hispanic	0.2%	26
Asian or Pacific Islander	0.4%	26
American Indian or Alaskan Native	0.1%	26
Race/ethnicity unknown	n/a	26
Nonresident alien	0.1%	26
Student Counts, 2001–2002:		
Full-time, first-time undergraduate	243	26
Percent full-time, first-time undergraduate	10.0%	26
Percent full-time, first-time undergraduate		
receiving financial aid	63.0%	26
Awards / Degrees Conferred (7/1/02–6/30/03)		
Associate degrees, total	172	27
Graduation Rates (cohort year 1999)		
Graduation rate within 150% of normal		
time to completion	19.6%	28
Transfer-out rate	9.2%	28
Graduation rates by gender		
Men	12.4%	28
Women	25.3%	28
Graduation rates by race/ethnicity		
White, non-Hispanic	19.5%	28
Black, non-Hispanic	n/a	28
Hispanic	n/a	28
Asian/Pacific Islander	n/a	28
American Indian/Alaska Native	n/a	28
Race/ethnicity unknown	n/a	28
Nonresident alien	n/a	28

Finance

Full-time tuition and fees costs (2003–2004)		
In-district	n/a	25
In-state	$1,581.00	25
Out-of-state	$5,708.00	25
REVENUES BY SOURCE, FY 2001		
Tuition and fees	$1,863,803	53
Federal appropriations	$0	53
State appropriations	$5,958,455	53
Local appropriations	$64,220	53
Federal grants and contracts	$2,858,227	53

	Data	Source (see Source List)
State grants and contracts	$413,096	53
Local grants and contracts	$0	53
Private gifts, grants, and contracts	$308,474	53
Endowment income	$0	53
Sales and services of educational activities	$2,400	53
Auxiliary enterprises	$930,332	53
Hospital revenues	$0	53
Other sources	$64,593	53
Independent operations	$0	53
Total current funds revenues	$12,463,600	53

EXPENDITURES BY FUNCTION, FY 2001		
Instruction	$5,226,669	53
Research	$0	53
Public service	$185,153	53
Academic support	$1,177,403	53
Student services	$977,505	53
Institutional support	$1,207,834	53
Operation and maintenance of plant	$682,410	53
Scholarships and fellowships	$1,955,080	53
Mandatory transfers	$0	53
Nonmandatory transfers	$500	53
Total educational and general expenditures	$11,412,554	53
Auxiliary enterprises	$891,476	53
Auxiliary enterprises (nonmandatory)	$64,018	53
Hospital expenditures	$0	53
Hospitals (nonmandatory)	$0	53
Independent operations	$0	53
Independent operations (nonmandatory)	$0	53
Other current funds expenditure	$0	53
Total current funds expenditures and transactions	$12,304,030	53
Amount salaries and wages, total E and G expenditures	$6,368,636	53
Employee fringe benefits, institutional	$1,436,585	53
E and G employee fringe benefits, paid from noninstitutional accounts	$0	53
Total E and G employee compensation	$7,805,221	53

Other Institutional Data

Geographic Location	Abingdon, VA	
Type of Institution	Public, 2-year	4
Degrees Awarded	Associate's	4
Carnegie Classification	Associate's College	4

Community Demographics

County	Washington	61
Population, 2000	51,405	61
Male	48.50%	61
Female	51.50%	61

Education Level		
County Percent High School Graduates, 2000	72.30%	61
County Percent Bachelor's or Higher, 2000	16.10%	61

	Data	Source (see Source List)
Race and Ethnicity, 2000		
White	97.60%	61
Black or African American	1.30%	61
American Indian and Alaska Native	0.10%	61
Asian	0.30%	61
Native Hawaiian and Other Pacific Islander	<0.10%	61
Persons reporting some other race	0.10%	61
Persons reporting two or more races	0.60%	61
Persons of Hispanic or Latino origin	0.60%	61
White persons, not of Hispanic/Latino origin	97.00%	61
Households, 2000	21,056	61
Persons per household, 2000	2.36	61
Median household income, 1999	$32,742	61
Per capita money income, 1999	$18,350	61
Persons below poverty, 1999	10.90%	61
Land Area, 2000 (square miles)	563.0	61
Persons per square mile, 2000	90.8	61
Nearest metropolitan area	Johnson City-Kingsport-Bristol, TN-VA	61
Employment		
Private nonfarm establishments		
with paid employees, 2001	1,174	61
Private nonfarm employment, 2001	14,499	61
Minority-owned firms, percent of total, 1997	<.10%	61
Women-owned firms, percent of total, 1997	22.20%	61

Piedmont Virginia Community College

	Data	Source (see Source List)

Student Demographics

Total Enrollment (Fall 2002)	4,273	30
Male	39.3%	30
Female	60.7%	30
White, non-Hispanic	82.8%	30
Black, non-Hispanic	12.2%	30
Hispanic	1.3%	30
Asian or Pacific Islander	2.1%	30
American Indian or Alaskan Native	0.4%	30
Race/ethnicity unknown	n/a	30
Nonresident alien	1.2%	30
Student Counts, 2001–2002:		
Full-time, first-time undergraduate	199	30
Percent full-time, first-time undergraduate	5.0%	30
Percent full-time, first-time undergraduate receiving financial aid	43.0%	30
Awards / Degrees Conferred (7/1/02–6/30/03)		
Associate degrees, total	225	31
Graduation Rates (cohort year 1999)		
Graduation rate within 150% of normal time to completion	15.2%	32
Transfer-out rate	13.5%	32
Graduation rates by gender		
Men	17.9%	32
Women	12.4%	32
Graduation rates by race/ethnicity		
White, non-Hispanic	17.1%	32
Black, non-Hispanic	insufficient data	32
Hispanic	n/a	32
Asian/Pacific Islander	n/a	32
American Indian/Alaska Native	n/a	32
Race/ethnicity unknown	n/a	32
Nonresident alien	n/a	32

Finance

Full-time tuition and fees costs (2003–2004)		
In-district	n/a	29
In-state	$1,922.00	29
Out-of-state	$6,346.00	29
REVENUES BY SOURCE, FY 2001		
Tuition and fees	$3,122,669	53
Federal appropriations	$0	53
State appropriations	$7,605,580	53
Local appropriations	$25,000	53
Federal grants and contracts	$2,075,923	53

	Data	Source (see Source List)
State grants and contracts	$258,915	53
Local grants and contracts	$0	53
Private gifts, grants, and contracts	$193,693	53
Endowment income	$0	53
Sales and services of educational activities	$14,579	53
Auxiliary enterprises	$98,089	53
Hospital revenues	$0	53
Other sources	$180,080	53
Independent operations	$0	53
Total current funds revenues	$13,574,528	53

EXPENDITURES BY FUNCTION, FY 2001

Instruction	$6,305,885	53
Research	$0	53
Public service	$274,432	53
Academic support	$1,957,661	53
Student services	$975,214	53
Institutional support	$1,636,692	53
Operation and maintenance of plant	$1,124,196	53
Scholarships and fellowships	$1,243,498	53
Mandatory transfers	$0	53
Nonmandatory transfers	$9,697	53
Total educational and general expenditures	$13,527,275	53
Auxiliary enterprises	$114,228	53
Auxiliary enterprises (nonmandatory)	$0	53
Hospital expenditures	$0	53
Hospitals (nonmandatory)	$0	53
Independent operations	$0	53
Independent operations (nonmandatory)	$0	53
Other current funds expenditure	$0	53
Total current funds expenditures and transactions	$13,641,503	53
Amount salaries and wages, total E and G expenditures	$7,781,810	53
Employee fringe benefits, institutional	$1,706,310	53
E and G employee fringe benefits, paid from noninstitutional accounts	$0	53
Total E and G employee compensation	$9,488,120	53

Other Institutional Data

Geographic Location	Charlottesville, VA	
Type of Institution	Public, 2-year	8
Degrees Awarded	Associate's	8
Carnegie Classification	Associate's College	8

Community Demographics

County	Albemarle	62
Population, 2000	87,670	62
Male	48.00%	62
Female	52.00%	62

Education Level

County Percent High School Graduates, 2000	87.40%	62
County Percent Bachelor's or Higher, 2000	47.70%	62

	Data	Source (see Source List)
Race and Ethnicity, 2000		
White	85.20%	62
Black or African American	9.70%	62
American Indian and Alaska Native	0.20%	62
Asian	2.90%	62
Native Hawaiian and Other Pacific Islander	<0.10%	62
Persons reporting some other race	0.90%	62
Persons reporting two or more races	1.30%	62
Persons of Hispanic or Latino origin	2.60%	62
White persons, not of Hispanic/Latino origin	83.60%	62
Households, 2000	31,876	62
Persons per household, 2000	2.44	62
Median household income, 1999	$50,749	62
Per capita money income, 1999	$28,852	62
Persons below poverty, 1999	6.70%	62
Land Area, 2000 (square miles)	723.0	62
Persons per square mile, 2000	109.7	62
Nearest metropolitan area	Charlottesville, VA	62
Employment		
Private nonfarm establishments		
with paid employees, 2001	1,763	62
Private nonfarm employment, 2001	22,062	62
Minority-owned firms, percent of total, 1997	5.40%	62
Women-owned firms, percent of total, 1997	26.10%	62

Community College of Denver

	Data	Source (see Source List)

Student Demographics

Total Enrollment (Fall 2002)	7,924	34
Male	37.8%	34
Female	62.2%	34
White, non-Hispanic	41.1%	34
Black, non-Hispanic	16.3%	34
Hispanic	32.0%	34
Asian or Pacific Islander	6.3%	34
American Indian or Alaskan Native	1.8%	34
Race/ethnicity unknown	n/a	34
Nonresident alien	2.7%	34
Student Counts, 2001–2002:		
Full-time, first-time undergraduate	501	34
Percent full-time, first-time undergraduate	8.0%	34
Percent full-time, first-time undergraduate		
receiving financial aid	59.0%	34
Awards / Degrees Conferred (7/1/02–6/30/03)		
Associate degrees, total	406	35
Graduation Rates (cohort year 1999)		
Graduation rate within 150% of normal		
time to completion	14.6%	36
Transfer-out rate	19.4%	36
Graduation rates by gender		
Men	8.5%	36
Women	19.1%	36
Graduation rates by race/ethnicity		
White, non-Hispanic	16.7%	36
Black, non-Hispanic	15.3%	36
Hispanic	12.8%	36
Asian/Pacific Islander	insufficient data	36
American Indian/Alaska Native	insufficient data	36
Race/ethnicity unknown	n/a	36
Nonresident alien	13.6%	36

Finance

Full-time tuition and fees costs (2003–2004)		
In-district	n/a	33
In-state	$1,889.00	33
Out-of-state	$8,558.00	33
REVENUES BY SOURCE, FY 2001		
Tuition and fees	$7,691,715	53
Federal appropriations	$0	53
State appropriations	$14,175,247	53
Local appropriations	$0	53
Federal grants and contracts	$8,001,491	53

	Data	Source (see Source List)
State grants and contracts	$1,793,928	53
Local grants and contracts	$94,953	53
Private gifts, grants, and contracts	$374,827	53
Endowment income	$0	53
Sales and services of educational activities	$763,253	53
Auxiliary enterprises	$974,744	53
Hospital revenues	$0	53
Other sources	-$1,338,786	53
Independent operations	$0	53
Total current funds revenues	$32,531,372	53

EXPENDITURES BY FUNCTION, FY 2001		
Instruction	$12,195,332	53
Research	$0	53
Public service	$597,835	53
Academic support	$3,493,547	53
Student services	$3,561,566	53
Institutional support	$2,906,487	53
Operation and maintenance of plant	$3,161,660	53
Scholarships and fellowships	$5,838,615	53
Mandatory transfers	$0	53
Nonmandatory transfers	$0	53
Total educational and general expenditures	$31,755,042	53
Auxiliary enterprises	$344,238	53
Auxiliary enterprises (nonmandatory)	$0	53
Hospital expenditures	$0	53
Hospitals (nonmandatory)	$0	53
Independent operations	$0	53
Independent operations (nonmandatory)	$0	53
Other current funds expenditure	$0	53
Total current funds expenditures and transactions	$32,099,280	53
Amount salaries and wages, total E and G expenditures	$15,511,102	53
Employee fringe benefits, institutional	$2,472,531	53
E and G employee fringe benefits, paid from noninstitutional accounts	$0	53
Total E and G employee compensation	$17,983,633	53

Other Institutional Data

Geographic Location	Denver, CO	
Type of Institution	Public, 2-year	12
Degrees Awarded	Associate's	12
Carnegie Classification	Associate's College	12

Community Demographics

County	Denver	63
Population, 2000	557,478	63
Male	50.50%	63
Female	49.50%	63

Education Level

County Percent High School Graduates, 2000	78.90%	63
County Percent Bachelor's or Higher, 2000	34.50%	63

	Data	Source (see Source List)
Race and Ethnicity, 2000		
White	65.30%	63
Black or African American	11.10%	63
American Indian and Alaska Native	1.30%	63
Asian	2.80%	63
Native Hawaiian and Other Pacific Islander	0.10%	63
Persons reporting some other race	15.60%	63
Persons reporting two or more races	3.70%	63
Persons of Hispanic or Latino origin	31.70%	63
White persons, not of Hispanic/Latino origin	51.90%	63
Households, 2000	239,235	63
Persons per household, 2000	2.27	63
Median household income, 1999	$39,500	63
Per capita money income, 1999	$24,101	63
Persons below poverty, 1999	14.30%	63
Land Area, 2000 (square miles)	153.0	63
Persons per square mile, 2000	3,616.8	63
Nearest metropolitan area	Denver, CO	63
Employment		
Private nonfarm establishments		
with paid employees, 2001	22,175	63
Private nonfarm employment, 2001	421,426	63
Minority-owned firms, percent of total, 1997	13.60%	63
Women-owned firms, percent of total, 1997	26.40%	63

Truman College

	Data	Source (see Source List)

Student Demographics

	Data	Source
Total Enrollment (Fall 2002)	15,280	38
Male	45.9%	38
Female	54.1%	38
White, non-Hispanic	22.5%	38
Black, non-Hispanic	13.6%	38
Hispanic	52.1%	38
Asian or Pacific Islander	11.3%	38
American Indian or Alaskan Native	0.5%	38
Race/ethnicity unknown	n/a	38
Nonresident alien	n/a	38
Student Counts, 2001–2002:		
Full-time, first-time undergraduate	368	38
Percent full-time, first-time undergraduate	2.0%	38
Percent full-time, first-time undergraduate receiving financial aid	69.0%	38
Awards / Degrees Conferred (7/1/02–6/30/03)		
Associate degrees, total	201	39
Graduation Rates (cohort year 1999)		
Graduation rate within 150% of normal time to completion	3.9%	40
Transfer-out rate	19.1%	40
Graduation rates by gender		
Men	1.7%	40
Women	5.7%	40
Graduation rates by race/ethnicity		
White, non-Hispanic	0.0%	40
Black, non-Hispanic	9.5%	40
Hispanic	2.7%	40
Asian/Pacific Islander	6.5%	40
American Indian/Alaska Native	n/a	40
Race/ethnicity unknown	n/a	40
Nonresident alien	n/a	40

Finance

	Data	Source
Full-time tuition and fees costs (2003–2004)		
In-district	$1,810.00	37
In-state	$5,485.00	37
Out-of-state	$7,879.00	37
REVENUES BY SOURCE, FY 2001		
Tuition and fees	$6,040,071	53
Federal appropriations	$0	53
State appropriations	$11,690,004	53
Local appropriations	$16,061,402	53
Federal grants and contracts	$6,592,698	53

	Data	Source (see Source List)
State grants and contracts	$8,331,971	53
Local grants and contracts	$0	53
Private gifts, grants, and contracts	$57,684	53
Endowment income	$0	53
Sales and services of educational activities	$70,956	53
Auxiliary enterprises	$0	53
Hospital revenues	$0	53
Other sources	$2,604,650	53
Independent operations	$0	53
Total current funds revenues	$51,449,436	53

EXPENDITURES BY FUNCTION, FY 2001		
Instruction	$17,941,815	53
Research	$0	53
Public service	$999,282	53
Academic support	$3,017,709	53
Student services	$2,725,064	53
Institutional support	$14,503,972	53
Operation and maintenance of plant	$3,621,968	53
Scholarships and fellowships	$4,255,662	53
Mandatory transfers	$628,134	53
Nonmandatory transfers	$891,240	53
Total educational and general expenditures	$48,584,846	53
Auxiliary enterprises	$0	53
Auxiliary enterprises (nonmandatory)	$0	53
Hospital expenditures	$0	53
Hospitals (nonmandatory)	$0	53
Independent operations	$0	53
Independent operations (nonmandatory)	$0	53
Other current funds expenditure	$0	53
Total current funds expenditures and transactions	$48,584,846	53
Amount salaries and wages, total E and G expenditures	$24,571,997	53
Employee fringe benefits, institutional	$0	53
E and G employee fringe benefits, paid from noninstitutional accounts	$7,552,247	53
Total E and G employee compensation	$32,124,244	53

Other Institutional Data

Geographic Location	Chicago, IL	
Type of Institution	Public, 2-year	16
Degrees Awarded	Associate's	16
Carnegie Classification	Associate's College	16

Community Demographics

County	Cook	64
Population, 2000	5,376,741	64
Male	48.40%	64
Female	51.60%	64

Education Level

County Percent High School Graduates, 2000	77.70%	64
County Percent Bachelor's or Higher, 2000	28.00%	64

	Data	Source (see Source List)
Race and Ethnicity, 2000		
White	56.30%	64
Black or African American	26.10%	64
American Indian and Alaska Native	0.30%	64
Asian	4.80%	64
Native Hawaiian and Other Pacific Islander	<.10%	64
Persons reporting some other race	9.90%	64
Persons reporting two or more races	2.50%	64
Persons of Hispanic or Latino origin	19.90%	64
White persons, not of Hispanic/Latino origin	47.60%	64
Households, 2000	1,974,181	64
Persons per household, 2000	2.68	64
Median household income, 1999	$45,922	64
Per capita money income, 1999	$23,227	64
Persons below poverty, 1999	13.50%	64
Land Area, 2000 (square miles)	946.0	64
Persons per square mile, 2000	5,685.6	64
Nearest metropolitan area	Chicago, IL	64
Employment		
Private nonfarm establishments		
with paid employees, 2001	127,162	64
Private nonfarm employment, 2001	2,515,882	64
Minority-owned firms, percent of total, 1997	20.10%	64
Women-owned firms, percent of total, 1997	26.40%	64

El Camino College

	Data	Source (see Source List)

Student Demographics

Total Enrollment (Fall 2002)	27,876	42
Male	46.7%	42
Female	53.3%	42
White, non-Hispanic	24.3%	42
Black, non-Hispanic	18.0%	42
Hispanic	28.3%	42
Asian or Pacific Islander	16.5%	42
American Indian or Alaskan Native	0.5%	42
Race/ethnicity unknown	9.8%	42
Nonresident alien	2.7%	42
Student Counts, 2001–2002:		
Full-time, first-time undergraduate	1281	42
Percent full-time, first-time undergraduate	5.0%	42
Percent full-time, first-time undergraduate		
receiving financial aid	29.0%	42
Awards / Degrees Conferred (7/1/02–6/30/03)		
Associate degrees, total	1021	43
Graduation Rates (cohort year 1999)		
Graduation rate within 150% of normal		
time to completion	38.6%	44
Transfer-out rate	16.1%	44
Graduation rates by gender		
Men	37.8%	44
Women	39.3%	44
Graduation rates by race/ethnicity		
White, non-Hispanic	49.5%	44
Black, non-Hispanic	21.8%	44
Hispanic	29.9%	44
Asian/Pacific Islander	46.4%	44
American Indian/Alaska Native	n/a	44
Race/ethnicity unknown	27.4%	44
Nonresident alien	53.7%	44

Finance

Full-time tuition and fees costs (2003–2004)		
In-district	n/a	41
In-state	$432.00	41
Out-of-state	$3,744.00	41
REVENUES BY SOURCE, FY 2001		
Tuition and fees	$5,653,921	53
Federal appropriations	$0	53
State appropriations	$42,784,974	53
Local appropriations	$0	53
Federal grants and contracts	$8,311,310	53

	Data	Source (see Source List)
State grants and contracts	$12,007,344	53
Local grants and contracts	$2,706,206	53
Private gifts, grants, and contracts	$0	53
Endowment income	$0	53
Sales and services of educational activities	$0	53
Auxiliary enterprises	$6,880,118	53
Hospital revenues	$0	53
Other sources	$0	53
Independent operations	$0	53
Total current funds revenues	$78,343,873	53

EXPENDITURES BY FUNCTION, FY 2001

Instruction	$38,635,711	53
Research	$490,017	53
Public service	$115,276	53
Academic support	$9,801,796	53
Student services	$12,079,263	53
Institutional support	$7,539,570	53
Operation and maintenance of plant	$6,847,882	53
Scholarships and fellowships	$7,338,653	53
Mandatory transfers	$2,438,913	53
Nonmandatory transfers	$100,000	53
Total educational and general expenditures	$85,387,081	53
Auxiliary enterprises	$3,246,451	53
Auxiliary enterprises (nonmandatory)	$100,000	53
Hospital expenditures	$0	53
Hospitals (nonmandatory)	$0	53
Independent operations	$0	53
Independent operations (nonmandatory)	$0	53
Other current funds expenditure	$0	53
Total current funds expenditures and transactions	$88,633,532	53
Amount salaries and wages, total E and G expenditures	$52,974,876	53
Employee fringe benefits, institutional	$8,593,963	53
E and G employee fringe benefits, paid from noninstitutional accounts	$0	53
Total E and G employee compensation	$61,568,839	53

Other Institutional Data

Geographic Location	Torrance, CA	
Type of Institution	Public, 2-year	20
Degrees Awarded	Associate's	20
Carnegie Classification	Associate's College	20

Community Demographics

County	Los Angeles	65
Population, 2000	9,871,506	65
Male	49.40%	65
Female	50.60%	65

Education Level

County Percent High School Graduates, 2000	69.90%	65
County Percent Bachelor's or Higher, 2000	24.90%	65

	Data	Source (see Source List)
Race and Ethnicity, 2000		
White	48.70%	65
Black or African American	9.80%	65
American Indian and Alaska Native	0.80%	65
Asian	11.90%	65
Native Hawaiian and Other Pacific Islander	0.30%	65
Persons reporting some other race	23.50%	65
Persons reporting two or more races	4.90%	65
Persons of Hispanic or Latino origin	44.60%	65
White persons, not of Hispanic/Latino origin	31.10%	65
Households, 2000	3,133,774	65
Persons per household, 2000	2.98	65
Median household income, 1999	$42,189	65
Per capita money income, 1999	$20,683	65
Persons below poverty, 1999	17.90%	65
Land Area, 2000 (square miles)	4,061.0	65
Persons per square mile, 2000	2,344.2	65
Nearest metropolitan area	Los Angeles-Long Beach, CA	65
Employment		
Private nonfarm establishments		
with paid employees, 2001	227,941	65
Private nonfarm employment, 2001	3,889,686	65
Minority-owned firms, percent of total, 1997	37.20%	65
Women-owned firms, percent of total, 1997	25.80%	65

Bakersfield College

	Data	Source (see Source List)

Student Demographics

Total Enrollment (Fall 2002)	16,138	46
Male	42.5%	46
Female	57.5%	46
White, non-Hispanic	45.1%	46
Black, non-Hispanic	7.1%	46
Hispanic	36.3%	46
Asian or Pacific Islander	5.6%	46
American Indian or Alaskan Native	1.9%	46
Race/ethnicity unknown	3.6%	46
Nonresident alien	0.5%	46
Student Counts, 2001–2002:		
Full-time, first-time undergraduate	1402	46
Percent full-time, first-time undergraduate	9.0%	46
Percent full-time, first-time undergraduate receiving financial aid	52.0%	46
Awards / Degrees Conferred (7/1/02–6/30/03)		
Associate degrees, total	868	47
Graduation Rates (cohort year 1999)		
Graduation rate within 150% of normal time to completion	34.6%	48
Transfer-out rate	23.3%	48
Graduation rates by gender		
Men	29.6%	48
Women	39.0%	48
Graduation rates by race/ethnicity		
White, non-Hispanic	35.9%	48
Black, non-Hispanic	11.1%	48
Hispanic	37.6%	48
Asian/Pacific Islander	insufficient data	48
American Indian/Alaska Native	n/a	48
Race/ethnicity unknown	insufficient data	48
Nonresident alien	insufficient data	48

Finance

Full-time tuition and fees costs (2003–2004)		
In-district	n/a	45
In-state	$538.00	45
Out-of-state	$4,710.00	45
REVENUES BY SOURCE, FY 2001		
Tuition and fees	$4,676,474	53
Federal appropriations	$268,520	53
State appropriations	$31,481,469	53
Local appropriations	$34,168,712	53
Federal grants and contracts	$16,129,595	53

	Data	Source (see Source List)
State grants and contracts	$20,407,141	53
Local grants and contracts	$2,722,810	53
Private gifts, grants, and contracts	$902,791	53
Endowment income	$0	53
Sales and services of educational activities	$76,845	53
Auxiliary enterprises	$6,829,122	53
Hospital revenues	$0	53
Other sources	$10,633,939	53
Independent operations	$0	53
Total current funds revenues	$128,297,418	53

EXPENDITURES BY FUNCTION, FY 2001

Instruction	$39,305,975	53
Research	$0	53
Public service	$338,947	53
Academic support	$10,855,812	53
Student services	$10,243,243	53
Institutional support	$25,320,989	53
Operation and maintenance of plant	$7,452,218	53
Scholarships and fellowships	$14,089,619	53
Mandatory transfers	$3,041,542	53
Nonmandatory transfers	$6,287,519	53
Total educational and general expenditures	$116,935,864	53
Auxiliary enterprises	$12,705,235	53
Auxiliary enterprises (nonmandatory)	$0	53
Hospital expenditures	$0	53
Hospitals (nonmandatory)	$0	53
Independent operations	$0	53
Independent operations (nonmandatory)	$0	53
Other current funds expenditure	$0	53
Total current funds expenditures and transactions	$129,641,099	53
Amount salaries and wages, total E and G expenditures	$58,467,678	53
Employee fringe benefits, institutional	$12,327,797	53
E and G employee fringe benefits, paid from noninstitutional accounts	$0	53
Total E and G employee compensation	$70,795,475	53

Other Institutional Data

Geographic Location	Bakersfield, CA	
Type of Institution	Public, 2-year	24
Degrees Awarded	Associate's	24
Carnegie Classification	Associate's College	24

Community Demographics

County	Kern	66
Population, 2000	713,087	66
Male	51.30%	66
Female	48.70%	66

Education Level

County Percent High School Graduates, 2000	68.50%	66
County Percent Bachelor's or Higher, 2000	13.50%	66

	Data	Source (see Source List)
Race and Ethnicity, 2000		
White	61.60%	66
Black or African American	6.00%	66
American Indian and Alaska Native	1.50%	66
Asian	3.40%	66
Native Hawaiian and Other Pacific Islander	0.10%	66
Persons reporting some other race	23.20%	66
Persons reporting two or more races	4.10%	66
Persons of Hispanic or Latino origin	38.40%	66
White persons, not of Hispanic/Latino origin	49.50%	66
Households, 2000	208,652	66
Persons per household, 2000	3.03	66
Median household income, 1999	$35,446	66
Per capita money income, 1999	$15,760	66
Persons below poverty, 1999	20.80%	66
Land Area, 2000 (square miles)	8,141.0	66
Persons per square mile, 2000	81.3	66
Nearest metropolitan area	Bakersfield, CA	66
Employment		
Private nonfarm establishments		
with paid employees, 2001	11,063	66
Private nonfarm employment, 2001	153,457	66
Minority-owned firms, percent of total, 1997	13.60%	66
Women-owned firms, percent of total, 1997	22.00%	66

Borough of Manhattan Community College

	Data	Source (see Source List)

Student Demographics

Total Enrollment (Fall 2002)	17,635	50
Male	36.4%	50
Female	63.6%	50
White, non-Hispanic	11.6%	50
Black, non-Hispanic	36.8%	50
Hispanic	29.4%	50
Asian or Pacific Islander	10.1%	50
American Indian or Alaskan Native	20.0%	50
Race/ethnicity unknown	n/a	50
Nonresident alien	11.9%	50

Student Counts, 2001–2002:

Full-time, first-time undergraduate	2686	50
Percent full-time, first-time undergraduate	17.0%	50
Percent full-time, first-time undergraduate receiving financial aid	70.0%	50

Awards / Degrees Conferred (7/1/02–6/30/03)

Associate degrees, total	2023	51

Graduation Rates (cohort year 1999)

Graduation rate within 150% of normal time to completion	9.7%	52
Transfer-out rate	12.6%	52

Graduation rates by gender

Men	7.7%	52
Women	11.3%	52

Graduation rates by race/ethnicity

White, non-Hispanic	17.0%	52
Black, non-Hispanic	9.2%	52
Hispanic	7.1%	52
Asian/Pacific Islander	5.3%	52
American Indian/Alaska Native	n/a	52
Race/ethnicity unknown	n/a	52
Nonresident alien	18.6%	52

Finance

Full-time tuition and fees costs (2003–2004)

In-district	n/a	49
In-state	$3,040.00	49
Out-of-state	$4,800.00	49

REVENUES BY SOURCE, FY 2001

Tuition and fees	$38,639,000	53
Federal appropriations	$0	53
State appropriations	$31,231,000	53
Local appropriations	$469,000	53
Federal grants and contracts	$23,086,666	53

	Data	Source (see Source List)
State grants and contracts	$11,249,334	53
Local grants and contracts	$912,414	53
Private gifts, grants, and contracts	$1,338,854	53
Endowment income	$0	53
Sales and services of educational activities	$0	53
Auxiliary enterprises	$650,000	53
Hospital revenues	$0	53
Other sources	$1,866,000	53
Independent operations	$0	53
Total current funds revenues	$109,442,268	53

EXPENDITURES BY FUNCTION, FY 2001

	Data	Source
Instruction	$37,980,000	53
Research	$110,251	53
Public service	$3,846,345	53
Academic support	$4,281,888	53
Student services	$7,993,159	53
Institutional support	$14,537,626	53
Operation and maintenance of plant	$8,215,000	53
Scholarships and fellowships	$30,757,000	53
Mandatory transfers	$61,000	53
Nonmandatory transfers	$64,000	53
Total educational and general expenditures	$107,846,269	53
Auxiliary enterprises	$744,000	53
Auxiliary enterprises (nonmandatory)	$0	53
Hospital expenditures	$0	53
Hospitals (nonmandatory)	$0	53
Independent operations	$0	53
Independent operations (nonmandatory)	$0	53
Other current funds expenditure	$0	53
Total current funds expenditures and transactions	$108,590,269	53
Amount salaries and wages, total E and G expenditures	$46,698,000	53
Employee fringe benefits, institutional	$11,222,000	53
E and G employee fringe benefits, paid from noninstitutional accounts	$0	53
Total E and G employee compensation	$57,920,000	53

Other Institutional Data

Geographic Location	New York, NY	
Type of Institution	Public, 2-year	28
Degrees Awarded	Associate's	28
Carnegie Classification	Associate's College	28

Community Demographics

County	New York	67
Population, 2000	1,537,195	67
Male	47.50%	67
Female	52.50%	67

Education Level

County Percent High School Graduates, 2000	78.70%	67
County Percent Bachelor's or Higher, 2000	49.40%	67

	Data	Source (see Source List)
Race and Ethnicity, 2000		
White	54.40%	67
Black or African American	17.40%	67
American Indian and Alaska Native	0.50%	67
Asian	9.40%	67
Native Hawaiian and Other Pacific Islander	0.10%	67
Persons reporting some other race	14.10%	67
Persons reporting two or more races	4.10%	67
Persons of Hispanic or Latino origin	27.20%	67
White persons, not of Hispanic/Latino origin	45.80%	67
Households, 2000	738644.00%	67
Persons per household, 2000	2.00	67
Median household income, 1999	$47,030	67
Per capita money income, 1999	$42,922	67
Persons below poverty, 1999	20.00%	67
Land Area, 2000 (square miles)	23.0	67
Persons per square mile, 2000	66,940.1	67
Nearest metropolitan area	New York, NY	67
Employment		
Private nonfarm establishments		
with paid employees, 2001	106,493	67
Private nonfarm employment, 2001	2,122,694	67
Minority-owned firms, percent of total, 1997	21.60%	67
Women-owned firms, percent of total, 1997	26.60%	67

Source List

1 *Wake Technical Community College: Institution detail.* (n.d.). Retrieved August, 2004, from National Center for Education Statistics Integrated Postsecondary Education Data System, College Opportunities Online site: http://nces.ed.gov/ipeds/cool/InstDetail.asp?UNITID=199856

2 *Wake Technical Community College: Enrollment.* (n.d.). Retrieved August, 2004, from National Center for Education Statistics Integrated Postsecondary Education Data System, College Opportunities Online site: http://nces.ed.gov/ipeds/cool/Enrollment.asp?UNITID=199856

3 *Wake Technical Community College: Awards and degrees conferred.* (n.d.). Retrieved August, 2004, from National Center for Education Statistics Integrated Postsecondary Education Data System, College Opportunities Online site: http://nces.ed.gov/ipeds/cool/Programs.asp? UNITID= 199856

4 *Wake Technical Community College: Graduation rates.* (n.d.). Retrieved August, 2004, from National Center for Education Statistics Integrated Postsecondary Education Data System, College Opportunities Online site: http://nces.ed.gov/ipeds/cool/GRS.asp?UNITID=199856

5 *Johnston Community College: Institution detail.* (n.d.). Retrieved August, 2004, from National Center for Education Statistics Integrated Postsecondary Education Data System, College Opportunities Online site: http://nces.ed.gov/ipeds/cool/InstDetail.asp?UNITID=198774

6 *Johnston Community College: Enrollment.* (n.d.). Retrieved August, 2004, from National Center for Education Statistics Integrated Postsecondary Education Data System, College Opportunities Online site: http://nces.ed.gov/ipeds/cool/Enrollment.asp?UNITID=198774

7 *Johnston Community College: Awards and degrees conferred.* (n.d.). Retrieved August, 2004, from National Center for Education Statistics Integrated Postsecondary Education Data System, College Opportunities Online site: http://nces.ed.gov/ipeds/cool/Programs.asp?UNITID=198774

8 *Johnston Community College: Graduation rates.* (n.d.). Retrieved August, 2004, from National Center for Education Statistics Integrated Postsecondary Education Data System, College Opportunities Online site: http://nces.ed.gov/ipeds/cool/GRS.asp?UNITID=198774

9 *Mountain View College: Institution detail.* (n.d.). Retrieved August, 2004, from National Center for Education Statistics Integrated Postsecondary Education Data System, College Opportunities Online site: http://nces.ed.gov/ipeds/cool/InstDetail.asp?UNITID=226930

10 *Mountain View College: Enrollment.* (n.d.). Retrieved August, 2004, from National Center for Education Statistics Integrated Postsecondary Education Data System, College Opportunities Online site: http://nces.ed.gov/ipeds/cool/Enrollment.asp?UNITID=226930

11 *Mountain View College: Awards and degrees conferred.* (n.d.). Retrieved August, 2004, from National Center for Education Statistics Integrated Postsecondary Education Data System, College Opportunities Online site: http://nces.ed.gov/ipeds/cool/Programs.asp?UNITID=226930

12 *Mountain View College: Graduation rates.* (n.d.). Retrieved August, 2004, from National Center for Education Statistics Integrated Postsecondary Education Data System, College Opportunities Online site: http://nces.ed.gov/ipeds/cool/GRS.asp?UNITID=226930

13 *Pima Community College: Institution detail.* (n.d.). Retrieved August, 2004, from National Center for Education Statistics Integrated Postsecondary Education Data System, College Opportunities Online site: http://nces.ed.gov/ipeds/cool/InstDetail.asp?UNITID=105525

14 *Pima Community College: Enrollment.* (n.d.). Retrieved August, 2004, from National Center for Education Statistics Integrated Postsecondary Education Data System, College Opportunities Online site: http://nces.ed.gov/ipeds/cool/Enrollment.asp?UNITID=105525

15 *Pima Community College: Awards and degrees conferred.* (n.d.). Retrieved August, 2004, from National Center for Education Statistics Integrated Postsecondary Education Data System, College Opportunities Online site: http://nces.ed.gov/ipeds/cool/Programs.asp?UNITID=105525

16 *Pima Community College: Graduation rates.* (n.d.). Retrieved August, 2004, from National Center for Education Statistics Integrated Postsecondary Education Data System, College Opportunities Online site: http://nces.ed.gov/ipeds/cool/GRS.asp?UNITID=105525

17 *GateWay Community College: Institution detail.* (n.d.). Retrieved August, 2004, from National Center for Education Statistics Integrated Postsecondary Education Data System, College Opportunities Online site: http://nces.ed.gov/ipeds/cool/InstDetail.asp?UNITID=105145

18 *GateWay Community College: Enrollme*nt. (n.d.). Retrieved August, 2004, from National Center for Education Statistics Integrated Postsecondary Education Data System, College Opportunities Online site: http://nces.ed.gov/ipeds/cool/Enrollment.asp?UNITID=105145

19 *GateWay Community College: Awards and degrees conferred.* (n.d.). Retrieved August, 2004, from National Center for Education Statistics Integrated Postsecondary Education Data System, College Opportunities Online site: http://nces.ed.gov/ipeds/cool/Programs.asp?UNITID=105145

20 *GateWay Community College: Graduation rates.* (n.d.). Retrieved August, 2004, from National Center for Education Statistics Integrated Postsecondary Education Data System, College Opportunities Online site: http://nces.ed.gov/ipeds/cool/GRS.asp?UNITID=105145

21 *Edmonds Community College: Institution detail.* (n.d.). Retrieved August, 2004, from National Center for Education Statistics Integrated Postsecondary Education Data System, College Opportunities Online site: http://nces.ed.gov/ipeds/cool/InstDetail.asp?UNITID=235103

22 *Edmonds Community College: Enrollment.* (n.d.). Retrieved August, 2004, from National Center for Education Statistics Integrated Postsecondary Education Data System, College Opportunities Online site: http://nces.ed.gov/ipeds/cool/Enrollment.asp?UNITID=235103

23 *Edmonds Community College: Awards and degrees conferred.* (n.d.). Retrieved August, 2004, from National Center for Education Statistics Integrated Postsecondary Education Data System, College Opportunities Online site: http://nces.ed.gov/ipeds/cool/Programs.asp?UNITID=235103

24 *Edmonds Community College: Graduation rates.* (n.d.). Retrieved August, 2004, from National Center for Education Statistics Integrated Postsecondary Education Data System, College Opportunities Online site: http://nces.ed.gov/ipeds/cool/GRS.asp?UNITID=235103

25 *Virginia Highlands Community College: Institution detail.* (n.d.). Retrieved August, 2004, from National Center for Education Statistics Integrated Postsecondary Education Data System, College Opportunities Online site: http://nces.ed.gov/ipeds/cool/InstDetail.asp?UNITID=233903

26 *Virginia Highlands Community College: Enrollment.* (n.d.). Retrieved August, 2004, from National Center for Education Statistics Integrated Postsecondary Education Data System, College Opportunities Online site: http://nces.ed.gov/ipeds/cool/Enrollment.asp?UNITID=233903

27 *Virginia Highlands Community College: Awards and degrees conferred.* (n.d.). Retrieved August, 2004, from National Center for Education Statistics Integrated Postsecondary Education Data System, College Opportunities Online site: http://nces.ed.gov/ipeds/cool/Programs.asp?UNITID=233903

28 *Virginia Highlands Community College: Graduation rates.* (n.d.). Retrieved August, 2004, from National Center for Education Statistics Integrated Postsecondary Education Data System, College Opportunities Online site: http://nces.ed.gov/ipeds/cool/GRS.asp?UNITID=233903

29 *Piedmont Virginia Community College: Institution detail.* (n.d.). Retrieved August, 2004, from National Center for Education Statistics Integrated Postsecondary Education Data System, College Opportunities Online site: http://nces.ed.gov/ipeds/cool/InstDetail.asp?UNITID=233116

30 *Piedmont Virginia Community College: Enrollment.* (n.d.). Retrieved August, 2004, from National Center for Education Statistics Integrated Postsecondary Education Data System, College Opportunities Online site: http://nces.ed.gov/ipeds/cool/Enrollment.asp?UNITID=233116

31 *Piedmont Virginia Community College: Awards and degrees conferred.* (n.d.). Retrieved August, 2004, from National Center for Education Statistics Integrated Postsecondary Education Data System, College Opportunities Online site: http://nces.ed.gov/ipeds/cool/Programs.asp?UNITID=233116

32 *Piedmont Virginia Community College: Graduation rates.* (n.d.). Retrieved August, 2004, from National Center for Education Statistics Integrated Postsecondary Education Data System, College Opportunities Online site: http://nces.ed.gov/ipeds/cool/GRS.asp?UNITID=233116

33 *Community College of Denver: Institution detail.* (n.d.). Retrieved August, 2004, from National Center for Education Statistics Integrated Postsecondary Education Data System, College Opportunities Online site: http://nces.ed.gov/ipeds/cool/InstDetail.asp?UNITID=126942

34 *Community College of Denver: Enrollment.* (n.d.). Retrieved August, 2004, from National Center for Education Statistics Integrated Postsecondary Education Data System, College Opportunities Online site: http://nces.ed.gov/ipeds/cool/Enrollment.asp?UNITID=126942

35 *Community College of Denver: Awards and degrees conferred.* (n.d.). Retrieved August, 2004, from National Center for Education Statistics Integrated Postsecondary Education Data System, College Opportunities Online site: http://nces.ed.gov/ipeds/cool/Programs.asp?UNITID=126942

36 *Community College of Denver: Graduation rates.* (n.d.). Retrieved August, 2004, from National Center for Education Statistics Integrated Postsecondary Education Data System, College Opportunities Online site: http://nces.ed.gov/ipeds/cool/GRS.asp?UNITID=126942

37 *City Colleges of Chicago, Harry S. Truman College: Institution detail.* (n.d.). Retrieved August, 2004, from National Center for Education Statistics Integrated Postsecondary Education Data System, College Opportunities Online site: http://nces.ed.gov/ipeds/cool/InstDetail.asp?UNITID=144184

38 *City Colleges of Chicago, Harry S. Truman College: Enrollment.* (n.d.). Retrieved August, 2004, from National Center for Education Statistics Integrated Postsecondary Education Data System, College Opportunities Online site: http://nces.ed.gov/ipeds/cool/Enrollment.asp?UNITID=144184

39 *City Colleges of Chicago, Harry S. Truman College: Awards and degrees conferred.* (n.d.). Retrieved August, 2004, from National Center for Education Statistics Integrated Postsecondary Education Data System, College Opportunities Online site: http://nces.ed.gov/ipeds/cool/Programs.asp?UNITID=144184

40 *City Colleges of Chicago, Harry S. Truman College: Graduation rates.* (n.d.). Retrieved August, 2004, from National Center for Education Statistics Integrated Postsecondary Education Data System, College Opportunities Online site: http://nces.ed.gov/ipeds/cool/GRS.asp?UNITID=144184

41 *El Camino College: Institution detail.* (n.d.). Retrieved August, 2004, from National Center for Education Statistics Integrated Postsecondary Education Data System, College Opportunities Online site: http://nces.ed.gov/ipeds/cool/InstDetail.asp?UNITID=113980

42 *El Camino College: Enrollment.* (n.d.). Retrieved August, 2004, from National Center for Education Statistics Integrated Postsecondary Education Data System, College Opportunities Online site: http://nces.ed.gov/ipeds/cool/Enrollment.asp?UNITID=113980

43 *El Camino College: Awards and degrees conferred.* (n.d.). Retrieved August, 2004, from National Center for Education Statistics Integrated Postsecondary Education Data System, College Opportunities Online site: http://nces.ed.gov/ipeds/cool/Programs.asp?UNITID=113980

44 *El Camino College: Graduation rates.* (n.d.). Retrieved August, 2004, from National Center for Education Statistics Integrated Postsecondary Education Data System, College Opportunities Online site: http://nces.ed.gov/ipeds/cool/GRS.asp?UNITID=113980

45 *Bakersfield College: Institution detail.* (n.d.). Retrieved August, 2004, from National Center for Education Statistics Integrated Postsecondary Education Data System, College Opportunities Online site: http://nces.ed.gov/ipeds/cool/InstDetail.asp?UNITID=109819

46 *Bakersfield College: Enrollment.* (n.d.). Retrieved August, 2004, from National Center for Education Statistics Integrated Postsecondary Education Data System, College Opportunities Online site: http://nces.ed.gov/ipeds/cool/Enrollment.asp?UNITID=109819

47 *Bakersfield College: Awards and degrees conferred.* (n.d.). Retrieved August, 2004, from National Center for Education Statistics Integrated Postsecondary Education Data System, College Opportunities Online site: http://nces.ed.gov/ipeds/cool/Programs.asp?UNITID=109819

48 *Bakersfield College: Graduation rates.* (n.d.). Retrieved August, 2004, from National Center for Education Statistics Integrated Postsecondary Education Data System, College Opportunities Online site: http://nces.ed.gov/ipeds/cool/GRS.asp?UNITID=109819

49 *CUNY Borough of Manhattan Community College: Institution detail.* (n.d.). Retrieved August, 2004, from National Center for Education Statistics Integrated Postsecondary Education Data System, College Opportunities Online site: http://nces.ed.gov/ipeds/cool/InstDetail.asp?UNITID=190521

50 *CUNY Borough of Manhattan Community College: Enrollment.* (n.d.). Retrieved August, 2004, from National Center for Education Statistics Integrated Postsecondary Education Data System, College Opportunities Online site: http://nces.ed.gov/ipeds/cool/Enrollment.asp?UNITID=190521

51 *CUNY Borough of Manhattan Community College: Awards and degrees conferred.* (n.d.). Retrieved August, 2004, from National Center for Education Statistics Integrated Postsecondary Education Data System, College Opportunities Online site: http://nces.ed.gov/ipeds/cool/Programs.asp?UNITID=190521

52 *CUNY Borough of Manhattan Community College: Graduation rates.* (n.d.). Retrieved August, 2004, from National Center for Education Statistics Integrated Postsecondary Education Data System, College Opportunities Online site: http://nces.ed.gov/ipeds/cool/GRS.asp?UNITID=190521

53 *NCES IPEDS Peer analysis tool.* (n.d.). Retrieved August, 2004, from National Center for Education Statistics Integrated Postsecondary Education Data System, Peer Analysis System site: http://nces.ed.gov/ipedspas/index.asp

54 *Wake County QuickFacts from the U.S. Census Bureau.* (n.d.). Retrieved August, 2004 from U.S. Census Bureau, State and County QuickFacts site: http://quickfacts.census.gov/qfd/states/37/37183.html

55 *Johnston County QuickFacts from the U.S. Census Bureau.* (n.d.). Retrieved August, 2004 from U.S. Census Bureau, State and County QuickFacts site: http://quickfacts.census.gov/qfd/states/37/37101.html

56 *Dallas County QuickFacts from the U.S. Census Bureau.* (n.d.). Retrieved August, 2004 from U.S. Census Bureau, State and County QuickFacts site: http://quickfacts.census.gov/qfd/states/48/48113.html

57 *Pima County QuickFacts from the U.S. Census Bureau.* (n.d.). Retrieved August, 2004 from U.S. Census Bureau, State and County QuickFacts site: http://quickfacts.census.gov/qfd/states/04/04019.html

58 *Maricopa County QuickFacts from the U.S. Census Bureau.* (n.d.). Retrieved August, 2004 from U.S. Census Bureau, State and County QuickFacts site: http://quickfacts.census.gov/qfd/states/04/04013.html

59 *Snohomish County QuickFacts from the U.S. Census Bureau.* (n.d.). Retrieved August, 2004 from U.S. Census Bureau, State and County QuickFacts site: http://quickfacts.census.gov/qfd/states/53/53061.html

60 *Washington County QuickFacts from the U.S. Census Bureau.* (n.d.). Retrieved August, 2004 from U.S. Census Bureau, State and County QuickFacts site: http://quickfacts.census.gov/qfd/states/51/51191.html

61 *Smyth County QuickFacts from the U.S. Census Bureau.* (n.d.). Retrieved August, 2004 from U.S. Census Bureau, State and County QuickFacts site: http://quickfacts.census.gov/qfd/states/51/51173.html

62 *Albemarle County QuickFacts from the U.S. Census Bureau.* (n.d.). Retrieved August, 2004 from U.S. Census Bureau, State and County QuickFacts site: http://quickfacts.census.gov/qfd/states/51/51003.html

63 *Denver County QuickFacts from the U.S. Census Bureau.* (n.d.). Retrieved August, 2004 from U.S. Census Bureau, State and County QuickFacts site: http://quickfacts.census.gov/qfd/states/08/08031.html

64 *Cook County QuickFacts from the U.S. Census Bureau.* (n.d.). Retrieved August, 2004 from U.S. Census Bureau, State and County QuickFacts site: http://quickfacts.census.gov/qfd/states/17/17031.html

65 *Los Angeles County QuickFacts from the U.S. Census Bureau.* (n.d.). Retrieved August, 2004 from U.S. Census Bureau, State and County QuickFacts site: http://quickfacts.census.gov/qfd/states/06/06037.html

66 *Kern County QuickFacts from the U.S. Census Bureau.* (n.d.). Retrieved August, 2004 from U.S. Census Bureau, State and County QuickFacts site: http://quickfacts.census.gov/qfd/states/06/06029.html

67 *New York County QuickFacts from the U.S. Census Bureau.* (n.d.). Retrieved August, 2004 from U.S. Census Bureau, State and County QuickFacts site: http://quickfacts.census.gov/qfd/states/36/36061.html

Appendix B

Participants in the Study: Institutions and Policy Officials

Institutions

Arizona: GateWay Community College; Pima Community College
California: Bakersfield College; El Camino College
Colorado: Community College of Denver
Illinois: Harry Truman College
New York: Borough of Manhattan Community College
North Carolina: Johnston Community College; Wake Technical Community College
Texas: Mountain View College
Virginia: Piedmont Community College; Virginia Highlands Community College
Washington: Edmonds Community College

Policy Officials

Arizona: Pima Community College Chancellor; President Pima College Community Campus; Maricopa District Community College Chancellor
California: Past and present Chancellors, California Community Colleges; President, Bakersfield Community College; Acting Chancellor, Kern Community College District
Colorado: President, Community College of Denver; Executive Director, Community Colleges of Colorado
New York: President, Borough of Manhattan Community College
North Carolina: President, North Carolina Community Colleges; President, Johnston Community College; President, Wake Technical Community College
Texas: Vice Chancellor, Associate Vice Chancellor, and Directors, Dallas Community College District
Washington: President, Edmonds Community College

Index